STRANGE &
UNEXPLAINED
HAPPENINGS

When Nature Breaks
the Rules of Science

STRANGE & UNEXPLAINED HAPPENINGS

When Nature Breaks the Rules of Science

volume 3

Jerome Clark and Nancy Pear

AN IMPRINT OF GALE RESEARCH,
AN INTERNATIONAL THOMPSON PUBLISHING COMPANY.

I(T)P

Changing the Way the World Learns

NEW YORK • LONDON • BONN • BOSTON • DETROIT • MADRID
MELBOURNE • MEXICO CITY • PARIS • SINGAPORE • TOKYO
TORONTO • WASHINGTON • ALBANY NY • BELMONT CA • CINCINNATI OH

STRANGE AND UNEXPLAINED HAPPENINGS:
When Nature Breaks the Rules of Science

Jerome Clark and Nancy Pear, *Editors*

STAFF

Sonia Benson, *U•X•L Developmental Editor*
Kathleen L. Witman, *U•X•L Associate Developmental Editor*
Carol DeKane Nagel, *U•X•L Managing Editor*
Thomas L. Romig, *U•X•L Publisher*

Margaret A. Chamberlain, *Permissions Associate (Pictures)*
Shanna P. Heilveil, *Production Associate*
Evi Seoud, *Assistant Production Manager*
Mary Beth Trimper, *Production Director*

Mary Krzewinski, *Art Director*
Cynthia Baldwin, *Product Design Manager*
Terry Colon, *Illustrator*

∞™ This book is printed on acid-free paper that meets the minimum requirements of American National Standard for Information Sciences—Permanence Paper for Printed Library Materials, ANSI Z39.48-1984.

ISBN 0-8103-9780-3 (Set)
  0-8103-9781-1 (Volume 1)
  0-8103-9782-X (Volume 2)
  0-8103-9889-3 (Volume 3)

Printed in the United States of America

I(T)P™

U•X•L is an imprint of Gale Research Inc., an International Thomson Publishing Company.
ITP logo is a trademark under license.

10 9 8 7 6 5 4 3 2 1

CONTENTS

VOLUME 1

I
UFOs: The Twentieth-Century Mystery 1

II
Ancient ETs and Their Calling Cards 23

VII
Light Shows 101

VIII
Strange Showers: Everything but Cats and Dogs 119

IX
More Weird Weather 147

VOLUME 2

X
Cryptozoology 191

XI
Misplaced Animals 201

XII
Shaggy, Two-footed Creatures in North America 229

XIX
Other Strange Events 487

READER'S GUIDE

Scope

Strange and Unexplained Happenings: When Nature Breaks the Rules of Science is a reliable reference guide to *physical* phenomena, as opposed to psychic or supernatural phenomena. It picks up where the studies of the occult and parapsychology leave off, telling the rest of the story of the world's mysteries—those dealing with strange natural and quasi-natural phenomena, things that seem to be a part of our world but are usually disputed or ignored by conventional science. No one who reads the newspaper or watches television is unfamiliar with reports of UFOs, the Loch Ness monsters, or Bigfoot. In *Strange* the stories of these and many more anomalies (abnormal or peculiar happenings) are presented with enough detail to stimulate wonder and surprise in even the most skeptical reader.

Research into such anomalies is not confined to a small group of nontraditional scientists. Academic teams, popular writers and members of anomalists' societies have examined strange physical phenomena in works ranging from sober, scientifically based analysis to wild conjecture. *Strange and Unexplained Happenings* presents the ideas of most of the major players in the field of each particular phenomenon, whether scientific or sensational. The entries also provide clear explanations of the kind of theory and research that have been applied. The greatest value of these fascinating accounts may not be in the answers they provide, but in the important questions they provoke.

Features

The three volumes of *Strange and Unexplained Happenings* are presented in chapters arranged by subject. Thus a student can look up Bigfoot and then browse the entire Shaggy, Two-footed Creatures in North America chapter, where he or she will find many other types of hairy bipeds that may or may not be related to the Bigfoot sightings. Similarly, by looking up "Flying Humanoids," the reader will find a variety of examples of visitors from other worlds.

Since anomalies, like most things in life, do not always fit neatly into one category or another, the volumes are extensively cross-referenced. Within each entry the names of related phenomena that have their own entries elsewhere in the set appear in boldface for quick reference. A thorough cumulative subject index concludes each volume.

The language used by anomalists (those who study strange happenings) presents a colorful variety of scientific and pseudoscientific terms. Although most of the terms are defined within the text, the glossary of terms appearing in the frontmatter of each volume ensures easy accessibility for all readers. Boxed material within the entry also provides interesting facts and explanations of concepts and terminology.

Strange centers on phenomena for which evidence is usually insufficient, if not altogether lacking. Photographs and drawings of all kinds have appeared as "proof" that some strange thing exists or happened. The photographs are often blurry and the drawings are often crude but, as the only evidence available, they are an important part of the history of the phenomenon and thus have been included in these volumes whenever possible.

Brief biographies of some of the foremost anomalists, whose works are most frequently cited within these volumes, are set off from the text in boxes. "Reel Life" boxes feature some modern and time-honored movies that have made such mysteries as werewolves, sea monsters, and UFOs a part of our culture.

Key sources are provided at the end of each entry. A Further Investigation section at the conclusion of each volume provides an annotated listing of the major books, periodicals, and organizations the student may wish to consult in further research into a particular strange phenomenon.

Comments and Suggestions

We welcome your comments on this work as well as your suggestions for strange events to be featured in future editions of *Strange and Unexplained Happenings*. Please write: Editors, *Strange and Unexplained Happenings*, U·X·L, 835 Penobscot Bldg., Detroit, Michigan 48226-4094; call toll-free: 1-800-877-4253; or fax: 313-961-6348.

INTRODUCTION

Strange and Unexplained Happenings: When Nature Breaks the Rules of Science is a book about *anomalies*, human experiences that go against common sense and break the rules that science uses to describe our world. In the words of folklorist Bill Ellis, "Weird stuff happens." *Strange and Unexplained Happenings* takes a look at the "weird stuff" that abounds in the reports of ordinary people who have had extraordinary experiences. Accounts of flying saucers, reptile men, werewolves, and abominable snowmen grab our attention and send shivers down our spines. When similar strange accounts are repeated time and again by different witnesses in different times and places, they capture the attention of the scientific community as well.

The three hardest words for human beings to utter are *I don't know*. Because we like our mysteries quickly and neatly explained, in modern times we have come to ask scientists to find logical explanations for strange human experiences. Sometimes science can use its knowledge and tools to find the answers to puzzling incidents; at other times it offers explanations that don't seem to fit the anomalies and only add to the confusion about them. When experiences are especially unbelievable, scientists may simply decide that they never really happened and refuse to consider them altogether. Most of us believe that as science learns more, it will be able to explain more. Still, it is almost certain that science will never be able to account for all the "weird stuff" that human beings encounter.

When an anomaly is reported, it is natural not to believe it, to be skeptical. One usually wonders about the witness. Could the person be lying for some reason? Tricks and hoaxes do occur. There are people who go to great lengths to fool scientists and the public, who hope to find fame and fortune by false claims or simply to prove to themselves how clever they are. Photographs of extraordinary happenings are often fake; it is thought that nearly 95 percent of all UFO photos are false, and some of the best film footage of the Loch Ness monster is judged doubtful as well. As a matter of fact, most investigators of anomalies feel that a lot of photographs of an incident signals a fake, because: (1) most people don't walk around with cameras ready to snap strange sights; (2) people having weird or scary experiences are often in shock or terror, and taking pictures is the last thing on their minds; and (3) anomalies generally last for just a matter of seconds. Investigators also believe that the fuzzier the photo, the more likely it is to be real, because pictures taken by people with shaking hands rarely turn out clearly, while hoaxers know that poor photographs won't get the results they are looking for.

It is also natural to wonder about the mental health of a person who has witnessed an extraordinary happening. Common sense tells us that *all* weird accounts should be blamed on the poor memories, bad dreams, or wild imaginings of confused and unwell minds! Still, psychologists who have examined witnesses of anomalies find them, for the most part, to be the same as people who have had no odd experiences at all. Also, the sheer number of strange reports rattles our common sense a bit, as do cases of multiple witnesses, when large groups of people observe the same strange happenings together.

More interesting still are accounts that have been repeated for centuries; reports of lake monsters in the deep waters of Loch Ness, for example, began way back in A.D. 565! Interesting, too, are reports that are widespread. The Pacific Northwest region of North America has its Bigfoot sightings, western Mongolians tell stories about the Almas, and accounts of the yeti have been reported in the high reaches of the Himalayas. While the languages and cultures surrounding these legends may differ, it is clear that witnesses are describing a similar creature: a hairy, two-legged "apeman." When observers report sightings of sea serpents they may describe them as smooth and snakelike or maned like horses, with many humps or finned like fish. Even when details vary widely, it is difficult to ignore their basic sameness: all suggest the existence of large, as yet unknown, sea-going animals.

It is true that in many cases of strange happenings, people have been misled or mistaken. The Bermuda Triangle, an area of the Caribbean where ships and planes were reported to mysteriously disappear, for example, was considered a real threat for more than two decades. That is, until weather records and other documents were properly researched, proving that the location was as safe as any other body of water. In the same way, the strange cattle "mutilations" that worried farmers in Minnesota and Kansas in the 1970s—and stirred up all sorts of weird explanations—required the special skills of veterinary pathologists to find that the cause was a simple, but gruesome, infection. Sometimes strange accounts do seem to change and grow as they are reported over the decades and in print. Human beings do want the truth ... but they also like a good story!

Science, too, has made its mistakes over the years. When sailors gave accounts of large sea creatures with giant eyes and many tentacles they were told that they were seeing floating trees with large roots. We now know that their accounts described giant squids. Gorillas and meteors were also rejected by scientists not that long ago!

But then, of course, some anomalies are more believable than others. When an odd happening turns the way we think about the world upside down it is described as "high-strange"; less weird accounts are lower on the strangeness scale. It is not *completely* unthinkable that unknown creatures still exist in some remote regions of the globe as many cryptozoologists (people who study "hidden" animals) believe. Wildlife experts and marine biologists may, over time, find that creatures like Bigfoot and "Nessie" are real. In the same way, physicists and meteorologists may find the reasons for ball lightning, or for the mysterious ice chunks that fall from the sky. The discovery of intelligent beings from outer space, on the other hand, would really shake us up and force us to rethink our lives and our place in the universe. As high on the strange scale as this idea is, though, there is enough hard evidence—like odd radar trackings and soil samples from UFO landing sights—to make it worth considering.

Strange accounts, no matter how farfetched, deserve some careful thought. Although most readers set their own limits as to how high on the strange scale they can go, the kinds of questions raised by anomalies are worth pursuing, even if the event or object is beyond one's own limits of belief. True understanding of anomalies takes time, effort, and an open—but not a gullible—mind. *Strange and Unexplained Happenings* doesn't deal with belief or disbelief. It only shows that human experiences come in more shapes and sizes than we could ever imagine!

GLOSSARY

A

anomaly: something that is abnormal and difficult to explain or classify by conventional systems. An *anomalist* is someone who studies or collects anomalies.

anthropoid: ape.

anthropology: the study of human beings in terms of their social relations with each other, their culture, where they live, where they come from, physical characteristics, and their relationship with the environment.

archaeology: the scientific study of prehistory by finding and examining the remains of past life, such as fossils, relics, artifacts, and monuments.

arkeology: a term used to describe the search for the remains of Noah's Ark at the site where it landed after the Great Flood, as chronicled in the Book of Genesis in the Bible.

astronomy: the study of things that are outside of the Earth's atmosphere.

Atlantis: a fabled island in the Atlantic inhabited by a highly advanced culture. According to Greek legend an earthquake caused the island to be swallowed up by the sea. Some still believe in the legendary Atlantis today, and societies have arisen in order to actively search for its remains.

atmospheric life forms: *See space animals.*

B

bioluminescent organisms: plants and animals that make their own light by changing chemical energy into light energy. Bioluminescent organisms are especially common in places where no light penetrates, like the depths of the ocean.

bipeds: animals that walk on two feet.

C

CE1: a UFO seen at less than 500 feet from the witness.

CE2: a UFO that physically affects its surroundings.

CE3: a being observed in connection with a UFO sighting.

cereology: the study of crop circles.

coelacanth: a large fish that, until 1938, had only been known through fossil records and was thought to have been extinct for some 60 million years. In 1938 a coelacanth was caught in the net of a South African fishing boat, giving rise to speculation that other species that had been officially declared extinct may live on.

contactee: a person who claims to have ongoing communications with one or more extraterrestrials. A *physical contactee* claims to have had actual physical contact with extraterrestrials and often will produce photographs or other material evidence of these meetings. A *psychic contactee* claims to have received messages from space in dreams or through automatic writing (writing performed without thinking, seemingly directed by an outside force).

corpse candles: also called death-candles; lights appearing in the form of a flame or a luminous mass, according to folk tradition, that foretell an impending death.

cover-up: an attempt made by an organization or group to conceal from the public the group's actions or information it has received or collected.

creationism: the belief, based on a word-for-word reading of the Bible's Book of Genesis, that God created all matter, all living things, and the world itself, all at the same time and from nothing.

creation myths: sacred stories that explain how the Earth and its beings were created.

Cro-Magnon race: a race that lived 35,000 years ago and is of the same species as modern human beings (*Homo sapiens*). Cro-Magnons stood straight and were six or more feet tall; their foreheads were high and their brains large. Skillfully made Cro-Magnon tools, jewelry, and cave wall paintings suggest that the Cro-Magnon race had an advanced culture.

cryptozoology: the study of lore concerning animals that science does not account for, including animals thought to be extinct, animals that have been seen only by local populations, or animals thought to exist only in certain areas that show up elsewhere. The objective of cryptozoology is generally to evaluate the possibility of these animals' existence.

D

debunk: to expose something as false or as a hoax.

dowsing: a folk method for finding underground water or minerals with a divining rod. The divining rod is usually a forked twig; the "diviner" holds the forked ends close to his or her body, and the stem supposedly points downward when he or she walks over the hidden water or desired mineral. Some believe that dowsing can be used as a method to predict when and where a crop circle will appear.

E

ethereans: fourth dimensional human beings; a theoretical group of beings like humans, only more advanced, who live in another (or fourth) dimension that coexists with our world. Just as the stars and planets of our universe have their etheric counterparts, ethereans are human beings in a different reality.

evolution: a process in which a group of plants or animals—such as a species—changes over a long period of time, so that descendants differ from their ancestors. Theoretically the changes result from *natural selection,* a process in which the strongest and the most adept at survival pass on their characteristics to the next generations. Characteristics that make group members less successful at surviving and breeding are slowly weeded out.

extinct: no longer in existence.

extraterrestrial: something that came into being or lives outside of the Earth's atmosphere, or something that happened outside of the Earth's atmosphere.

F

Fortean: an adjective used to describe outlandish, sometimes sarcastic, and generally antiscientific theories in the literature of the strange and unexplained. The word is derived from the pioneer of physical anomalies, Charles Fort, who frequently poked fun at the weak attempts science made to explain away strange events by offering wacky theories of his own.

G

geophysics: a branch of earth science dealing with physical processes and phenomena occurring within or on the Earth.

H

hallucination: an illusion of seeing, hearing, or in some way becoming aware of something that apparently does not exist in reality.

herpetology: the scientific study of reptiles and amphibians.

humanoid: having human characteristics; a being that resembles a human.

I

ichthyology: the study of fish.

inorganic: composed of matter other than plant or animal; relating to mineral matter as opposed to the substance of things that are or were alive.

L

Lemuria: a legendary lost continent in the Pacific Ocean somewhere between southern Africa and southern India. Unlike accounts of Atlantis, which date back to the writings of Plato in ancient Greece, theories about Lemuria arose in the nineteenth century in the doctrine of occultists such as Madame Helena Petrovna Blavatsky, cofounder of the Theosophical Society, Max Heindel, founder of the Rosicrucian Fellowship, Rudolf Steiner, founder of the Anthroposophical Society, and Theosophist W. Scott-Elliott.

lycanthropy: the transformation of a man or woman into a wolf or wolflike human.

M

meteor: one of the small pieces of matter in the solar system that can be seen only when it falls into the Earth's atmosphere, where friction may cause it to burn or glow. When this happens it is sometimes called a "falling" or "shooting" star.

meteorite: a meteor that survives the fall to Earth.

meteoroid: any piece of matter—ranging in mass from a speck of dust to thousands of tons—that travels through space; it is composed largely of stone or iron or a mixture of the two. When a meteoroid enters the Earth's atmosphere it becomes visible and is called a *meteor.*

meteorology: the science of weather and other atmospheric phenomena.

mollusks: the second largest group of invertebrate animals (those without a backbone). They are soft-bodied, and most have a distinct shell. Mollusks usually live in water and include scallops, clams, oysters, mussels, snails, squids, and octopuses.

mutology: the investigation of cattle mutilations.

N

Neanderthal race: a species that lived between 40,000 and 100,000 years ago. Neanderthal remains have been found in Europe, northern Africa, the Middle East, and Siberia. The classic Neanderthal man had a large thick skull with heavy brow ridges, a sloping forehead, and a chinless jaw. He was slightly over five feet tall and had a stocky body. The link between Neanderthal man and modern human beings is unclear. Many anthropologists believe that they evolved separately from an earlier common ancestor.

New Age: a late-twentieth-century social movement that draws from American Indian and Eastern traditions and espouses spirituality, holism, concern for the environment, and metaphysics.

O

occultism: belief in or study of supernatural powers.

OINTS (Other Intelligences): a term coined by biologist Ivan T. Sanderson that includes not only extraterrestrials, but also space animals, undersea civilizations, poltergeists, and extradimensional beings.

organic: composed of living plant or animal matter.

ornithology: the scientific study of birds.

P

paleontology: the scientific study of the past through fossils and ancient forms of life.

paracryptozoology: the study of animals whose existence, even to the most open-minded, seems impossible (*paracryptozoology* means "beyond cryptozoology").

paranormal: not scientifically explainable; supernatural.

paraphysical: a combination of the terms *paranormal* (outside the normal) and *physical*. This concept, used by some anomalists, encompasses both natural occurrences (like leaving tracks) and unnatural occurrences (like disappearing instantly).

plesiosaurs: a suborder of prehistoric reptiles that dominated the seas during the Cretaceous Period (136 to 65 million years ago). Their bodies were short, broad, and flat. They had short pointed tails. Their small heads were supported by long slender necks, ideal for darting into the water to catch fish. Plesiosaurs swam with a rowing movement, using their four powerful, diamond-shaped flippers like paddles. They were often quite large, measuring up to 40 feet in length.

porphyria: a rare genetic disease often linked with werewolf sightings. Porphyria sufferers are plagued by tissue destruction in the face and fingers, open sores, and extreme sensitivity to light. Their facial skin may take on a brownish cast, and they may also suffer from mental illness. The inability to tolerate light, plus shame stemming from physical deformities, may lead the afflicted to venture out only at night. Some in the medical community have suggested that sightings of werewolves have really been of individuals with porphyria.

primates: a member of the group of mammals that includes man, apes, monkeys, and prosimians, or lower primates. Primates have highly developed brains and hands with opposable thumbs that are very adept at holding and grasping things.

psychosocial hypothesis: a belief that UFOs and other anomalies are powerful hallucinations shaped by the witness's psyche and culture, and that strange sightings are actually insights into deep realms of the human imagination rather than evidence of visitors from other worlds.

Q

quadrupeds: animals that walk on four feet.

R

radar: a method of detecting distant objects and determining their position, speed, or other characteristics by analyzing radio waves reflected from them.

S

saucer nests: circular indentations that one could imagine to be left by hovering or grounded spacecraft; saucer nests, found in the 1960s and 1970s before the current crop circle mania began, have many features comparable to crop circles, but the connection between the two phenomena has not been established.

sauropods: huge plant-eating reptiles with long necks and tails, small heads, bulky bodies, and stumplike legs; *Diplodocus, Apatosaurus (Brontosaurus),* and *Brachiosaurus* were sauropods.

shape-shifter (or shape-changer): one who can change form at will; in medieval and later chronicles, shape-shifting was associated with witchcraft, and such shape-shifters as black dogs and werewolves were often considered to be either agents of the devil or Satan himself.

sonar: a method of tracking that uses reflected sound waves to detect and locate underwater objects.

space animals (atmospheric life forms): hypothetical life forms existing in the upper atmosphere. Several ufologists have suggested that UFOs are neither spacecraft nor cases of mistaken identity but *space animals.*

spontaneous generation: a once widespread belief that living things can spring from nonliving material; thus, when rain hits the ground it can give rise—out of the mud, slime, and dust—to all sorts of living matter.

supernatural: of or relating to an order of existence outside the natural, observable universe.

T

teleportation: the act of moving an object or person from one place to another by using the mind, without using physical means.

theosophy: teachings about God and the world based on mystical insight; in 1875 the Theosophy movement arose in the United States, following Eastern theories of evolution and reincarnation.

transient lunar phenomena (TLP): unusual, short-lasting appearances on the moon's surface typically observed by astronomers through telescopes, and more rarely by the naked eye.

U

UFO (unidentified flying object): a term first coined by a U.S. Air Force worker that came into common usage in the mid-1950s to describe the "flying saucers" or mysterious discs that were being observed in the air and were suspected by some to be the craft of extraterrestrial visitors.

UFO-abduction reports: the accounts of a significant group of witnesses who claim to have been kidnapped by aliens. Many witnesses described large-headed, gray-skinned humanoids who subjected them to medical examinations. Some witnesses experienced amnesia after their encounters and recalled them only through hypnosis.

ufology: the study of unidentified flying objects.

ultraterrestrials: beings from another reality.

W

water horse: a folkloric creature believed by many to be a dangerous shape-changer that can appear either as a shaggy man who leaps out of the dark onto the back of a lone traveler or as a young horse that, after tricking an unknowing soul onto its back, plunges to the bottom of the nearest lake, killing its rider.

waterspouts: funnel- or tube-shaped columns of rotating, cloud-filled wind, usually extending from a cloud down to the spray it tears up from the surface of an ocean or lake.

Z

zeuglodon: a primitive, snakelike whale thought to have become extinct long ago.

PICTURE CREDITS

The photographs and illustrations appearing in *Strange and Unexplained Happenings* were received from the following sources:

Fortean Picture Library: pp. 6, 7, 26, 44, 46, 48, 82, 117, 122, 129, 130, 139, 158, 210, 212, 216, 258, 267, 274, 282, 285, 296, 332, 341, 346, 381, 383, 389, 396, 399, 403, 412, 420, 421, 427, 429, 432, 435, 450, 468, 482; **J. Allen Hynek Center for UFO Studies, collection of Jerome Clark:** p. 8; **Archive Photos/Lambert:** p. 28; **Archive Photos:** pp. 29, 65, 68, 79, 232, 266, 314, 391; **Loren Coleman:** pp. 32, 196, 197, 204, 227, 233, 270, 312, 316, 331, 377, 392, 441, 463, 494, 509; **The Bettmann Archive:** pp. 61, 256, 283, 323, 449; **The Granger Collection:** pp. 67, 276, 385, 395, 417; **UPI/Bettmann:** pp. 91, 93, 152, 154, 161, 164, 194, 206, 235, 246, 250, 424, 447; **AP/Worldwide Photos:** p. 95; **Drawing by Don Schmitt, collection of Jerome Clark:** p. 113; **Archive Photos/Shelly Grossman:** p. 115; *Fort Worth Star-Telegram:* p. 127; **Western Mail & Echo, Cardiff:** p. 138; **Sovfoto/Eastfoto:** p. 163; **Loren Coleman/Jim McClarin:** p. 242; **Rene Dahinden:** p. 245; **Loren Coleman/George Holton:** p. 269; **Archive Photos/D.P.A.:** p. 279; **Marie T. Womack from** *A Living Dinosaur?: In Search of Mokele-Mbembe* **by Dr. Roy P. Mackal ©1987 by E. J. Brill, Leiden, The Netherlands:** p. 307; **Archive Photos/Max Hunn:** p. 336; **Archive Photos/Tamara Andreeva:** p. 342; **Archive Photos/Herbert:** p. 443; **Courtesy of Dennis Stacy:** pp. 458, 500, 501; **University of Leeds Library:** pp. 460, 461; **U.S. Fish and Wildlife Service photo by Curtis Carley:** p. 471; **U.S. Fish and Wildlife Service photo by Glen Smart:** p. 479; **Mary Evans Picture Library:** p. 490; **Larry E. Arnold:** p. 492.

STRANGE &
UNEXPLAINED
HAPPENINGS

When Nature Breaks
the Rules of Science

Monsters of
the Deep

- UNIDENTIFIED SUBMARINE OBJECTS

- GIANT OCTOPUS

- GIANT SQUID

- SEA SERPENTS

Monsters of the Deep

UNIDENTIFIED SUBMARINE OBJECTS

While sailing in the Atlantic Ocean near the equator in the early morning hours of October 28, 1902, the *Fort Salisbury* came upon an incredible sight. In the ship's log, Second Officer A. H. Raymer noted a "dark object, with long, luminous trailing wake" in a "phosphorescent sea." On the object were what appeared to be two bright "masthead" or "steamer's lights." Still, Raymer thought the dark mass was some kind of bioluminescent whale, for it sank below the surface of the water as they drew nearer to it.

Approaching the watery trail it left behind, Raymer again came upon the object, or rather, its scaly back as it slowly lowered itself in the water. It was still too early in the morning for a clear look, but the scales appeared to be a foot across and "dotted in places with barnacle growth." The thing was some 30 feet across at its widest point, tapering "at the extreme end." It was "about 500 ft. to 600 ft." in length.

Raymer noted that "the monster's progress could be distinctly heard." He also reported that "the wet shiny back of the monster was dotted with twinkling phosphorescent lights, and was encircled with a band of white phosphorescent sea." A helmsman and a lookout also witnessed the sight.

This account was reported in the *London Daily Mail* of November 9, 1902. When interviewed, the ship's captain—who was not himself a witness—remarked of his second officer, "I can only

BIOLUMINESCENT ORGANISMS

Bioluminescent organisms are plants and animals that make their own light. They do this by changing chemical energy into light energy. Different plants and animals produce different bioluminescent chemicals. Bioluminescent organisms are especially common in places where no light penetrates, like the depths of the ocean.

say that he is very earnest on the subject and has, together with the lookout and the helmsman, seen something in the water, of a huge nature, as specified."

Demon of the Deep

An even more fantastic story was published in a Washington newspaper, the *Tacoma Daily Ledger*, on July 3, 1893. The witnesses were members of a fishing party that left Tacoma on the afternoon of July 1 and were camping that evening on Henderson Island, not far from a large group of surveyors. One of the fishermen related that he was startled awake some time after midnight (though he could not be certain because his watch and all others belonging to the group had mysteriously stopped) by "a most horrible noise [that] rang out in the clear morning air, and instantly the whole air was filled with a strong current of electricity that caused every nerve in the body to sting with pain." A bright light also flashed constantly; "at first I thought it was a thunderstorm," the witness said, "but as no rain accompanied it, and as both light and sound came off the bay, I turned my head in that direction." What he saw was "a most horrible–looking monster."

By now, according to the witness, every man in the fishing camp, as well as the nearby surveyors, had "gathered on the bank of the stream.... The monster slowly drew in toward the shore, and as it approached, from its head poured out a stream of water that looked like blue fire. All the while the air seemed to be filled with electricity, and the sensation experienced was as if each man had on a suit of clothes formed of the fine points of needles."

One of the surveyors foolishly took a few steps forward and some of the water being shot from the monster's head reached him; "he instantly fell to the ground and lay as though dead." A second man tried to pull the fellow's body to safety, but he was also struck by the water and met the same fate. That was enough for the rest of the group! They rushed into the woods in a panic.

But even in the woods, the "demon of the deep," as the witness called it, "sent out flashes of light that illuminated the surrounding country for miles, and his roar—which sounded like the roar of thunder—became terrific." But the monster then changed direction, "and in an instant ... disappeared beneath the waters of the bay," though those present "for some time ... were able to trace its course by a bright luminous light that was on the surface of the water." Afterward, left in total darkness, the men struggled to find their way back to the bay.

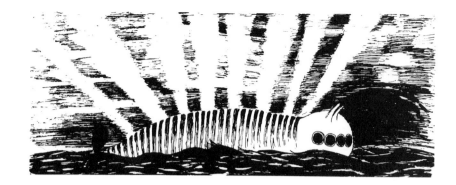

Fortunately their fallen companions were unconscious, not dead. As all awaited daybreak, the two revived.

As the witness described the "monster fish," it was "150 feet long" and 30 feet around "at its thickest part." It had a head shaped like that of a walrus, with six dull eyes the size of dinner plates. "At intervals of about every eight feet from its head to its tail a substance that had the appearance of a copper band encircled its body, and it was from these many bands that the powerful electric current appeared to come," the witness said. "Near the center of its head were two hornlike substances, ... it was through them that the electrically charged water was thrown." Stranger still, "its tail ... was shaped like a propeller and seemed to revolve."

Other USOs

But not all reports of unidentified submarine objects (USOs) are quite so fantastic. Even in the nineteenth century, some surprisingly believable accounts were reported. One such example occurred on June 18, 1845, in the eastern Mediterranean Sea. The crew of the brig *Victoria* saw three bright, glowing objects emerge from the water and shoot into the sky, where they were visible for ten minutes.

Professor Baden–Powell collected testimony from the crew and other witnesses and published it in an 1861 issue of *Report of the British Association*. One man who observed the sight from land said the objects were larger than the moon and had things like sails or streamers trailing from them. He and other witnesses, who had watched the objects for 20 minutes to an hour, said they appeared to be joined together.

On November 12, 1887, a "large ball of fire" ("enormous" by one account) rose from the sea near Cape Race, Newfoundland, Canada.

When it was 50 feet in the air it approached a nearby ship, the British steamer *Siberian,* moving against the wind as it did. It then retreated and flew away. The event was discussed in *Nature, L'Astronomie,* and *Meteorological Journal.*

There have been a number of modern USO reports as well; most have been associated with UFOs. But some of these "otherworldly" sightings may have been of earthly submarines in unexpected places. During the 1960s and 1970s, for example, Scandinavia was beset by submarines in its surrounding waters, and some writers claimed that the vessels were of mysterious origin. They were, in fact, Soviet craft on spy missions.

Below are some of the more puzzling modern USO reports:

Off the Alaska coast, summer 1945

Around sunset, a large round object, some 200 feet across, emerged from the sea about a mile from the U.S. Army transport Delarof. After rising in the air a short distance, it approached the ship, circling it silently two or three times before it flew off and disappeared into the southwest.

North Atlantic, late summer 1954

The crew of the Dutch ship Groote Beer observed a flat, moon-shaped object rise out of the ocean. Watching it through binoculars, Captain Jan P. Boshoff noted that it was gray and ringed by bright lights. It flew off with amazing speed. That same night it or another unknown object was seen by the Honduran freighter Aliki P., which was sailing in the same general area. It radioed the Long Island Coast Guard: "Observed ball of fire moving in and out of water without being extinguished. Trailing white smoke. Moving in erratic course, finally disappeared."

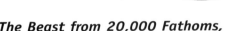

REEL LIFE

The Beast from 20,000 Fathoms, 1953.

Atomic testing defrosts a giant dinosaur in the Arctic; the hungry monster proceeds onward to its former breeding grounds, now New York City. The saurian-on-the-loose formula is fun, and good special effects bring it to life.

Octaman, 1971.

Comical thriller featuring non-threatening octopus-man discovered by scientists in Mexico. By the director of *The Creature of the Black Lagoon.*

20,000 Leagues Under the Sea, 1954.

From a futuristic submarine, Captain Nemo wages war on the surface world. A shipwrecked scientist and sailor do their best to thwart Nemo's dastardly schemes. Buoyant Disney version of French writer Jules Verne's 1870 novel.

Westchester County, New York, September 17, 1955

Around 1:30 A.M.., a couple named Bordes was rowing out onto a lake on Titicus Reservoir to fish when they saw an object come out of the water a few feet from their boat. Rose-colored and glowing, it was the size and shape of a basketball. After rising a foot in the air, it fell back into the water with a loud splash and disappeared. Frightened, the Bordeses headed for shore. On the way, they saw two wavy parallel lights, 30 feet long, on or just below the surface of the water at the center of the lake. A round, yellowish-white light hung above them. This light acted like a rotating spotlight; it appeared to be attached to what looked like a dim gray form in the water. The couple watched the object for a time and even tried to approach it; when they left it was still there. The couple were certain that they had not seen another boat.

Shag Harbor, Nova Scotia, October 4, 1967

Just before midnight, two men driving in a car saw a row of bright reddish-orange lights. They "came off and on one at a time," according to Nova Scotia's Yarmouth Light Herald (October 12). Five other people in a car stopped and watched the lights fly off to the water, where they changed into a single bright white light that bobbed on the waves. Many others, including a Royal Canadian Mounted Police officer, also saw the light, and a number of boats set out to find it. "Within an hour," the paper related, "the boats had arrived in the area where the object had disappeared, and reported finding a very large patch of bubbling water and foam. One fisherman described the froth as 80 feet wide and yellowish in color and said that he had never seen anything like it before in the area."

"Within an hour," the paper related, "the boats had arrived in the area where the object had disappeared, and reported finding a very large patch of bubbling water and foam."

Underwater civilization?

Taking reports like these and combining them with **Bermuda Triangle** disappearance stories, biologist Ivan T. Sanderson wrote a book, *Invisible Residents* (1970), about an advanced underwater civilization of OINTS (Other Intelligences) that were sometimes forced to seize planes and ships to keep their presence secret. "If a superior technological type of intelligent civilization(s) developed on this planet under water," he wrote, "they would very likely have gotten much farther ahead than we have, having had several millions, and possibly up to a billion years' headstart on us, life as we know it having started in the sea."

While Sanderson had no real evidence for his extraordinary ideas, it has been suggested from time to time that space beings maintain bases in oceans and lakes, and that their craft are exploring these watery regions as they allegedly explore the earth's land areas.

Sources:

Coleman, Loren, *Curious Encounters: Phantom Trains, Spooky Spots, and Other Mysterious Wonders,* Boston: Faber and Faber, 1985.
Heuvelmans, Bernard, *In the Wake of the Sea-Serpents,* New York: Hill and Wang, 1968.
Sanderson, Ivan T., *Invisible Residents: A Disquisition upon Certain Matters Maritime, and the Possibility of Intelligent Life Under the Waters of This Earth,* New York: World Publishing Company, 1970.

GIANT OCTOPUS

Bicycling on Anastasia Island, Florida, on the evening of November 30, 1896, Saint Augustine residents Herbert Coles and Dunham Coretter found the immense remains of a water animal on the beach. Because of its great weight, the huge carcass had sunk far into the sand when Coles and Coretter spotted it. They did not take measurements, but they knew right away that it was bigger than anything they had ever heard of.

The next day DeWitt Webb, a physician and founder of the St. Augustine Historical Society and Institute of Science, went to the site with several associates. The group concluded that the creature, which appeared to have been beached just days before, weighed nearly five tons. The parts that were visible measured 23 feet in length, four feet high, and 18 feet across the widest part of the back. The skin was a faint pink that seemed almost white and had a silvery cast to it. It was not a whale, Webb decided. It could only be some kind of octopus—of a size never before imagined.

Webb and his assistants returned to the beach as time and weather allowed over the next few days and took photographs—since lost—of the decayed, mutilated remains. One assistant, alone on one trip, reportedly found large pieces of arms while digging near the carcass. According to an account in the April 1897 issue of *American Naturalist,* "One arm was lying west of the body, 23 feet long; one stump of arm, west of body, about four feet; three arms lying south of body and from appear-

A common octopus.

ance attached ... longest one measured over 32 feet, the other arms were three to five feet shorter." It appeared that the animal had been attacked and partially torn apart before its body had washed ashore.

Soon afterward a strong storm arose, and the carcass floated out to sea. It washed back to shore two miles south of the original site.

A Case of Mistaken Identity

Webb began writing letters to scientists he thought would be interested in the find. One, dated December 8, 1896, made its way to A. E. Verrill, a Yale University zoologist known for his pioneering work on the once legendary—but now recognized—giant squid or Kraken. Verrill disagreed with Webb's suggestion that the carcass was of an octopus, for the largest known specimen of its kind measured 25 feet. He thought that the beached creature was a giant squid and said so briefly in the January 1897 issue of the *American Journal of Science.* But with further information, he accepted the giant octopus identification.

From comparing the arm pieces of the beached animal to those of known octopus specimens, he reached a fantastic conclusion: its full arm length must have been at least 75 feet! Truly monstrous in size, it would have measured 200 feet from tentacle tip to tentacle tip. Though Webb had been most instrumental in bringing the find to the attention of the scientific community, Verrill named the new animal after himself: *Octopus giganteus Verrill.*

Meanwhile, stormy weather conditions had moved the carcass once again. By the time it settled in its third location, even more of the body was missing. Regardless, what remained was still too heavy to move. Finally, in a January 17, 1897, letter to W. H. Dall, curator of mollusks at the National Museum in Washington, D. C., Webb related how he had had some luck rolling the creature out of its pit along some heavy planking with the help of four horses and six men. Now resting on planks 40 feet higher on the beach, the specimen was safe from washing away once more.

In the letter to Dall, Webb elaborated on the specimen:

A good part of the mantle or head remains attached near to the more slender part of the body.... The body was then opened for the entire length of 21 feet.... The slender part of the body was

---

MOLLUSKS

Mollusks make up the second-largest group of invertebrate animals (those without a backbone). They are soft-bodied, and most have a distinct shell. Mollusks usually live in water and include scallops, clams, oysters, mussels, snails, squids, and octopuses.

Fisherman attacked
by octopus; from a
drawing by Hokusai,
a celebrated Japanese
artist in the early
twentieth century.

entirely empty of internal organs. And the organs of the remainder were not large and did not look as if the animal had been long dead.... The muscular coat which seems to be all there is of the invertebrate is from two and three to six inches in thickness.

He noted no tailfin or other fins, "no beak·or head or eyes remaining," and "no pen to be found nor any evidence of any bony structure whatever." (The "pen" Webb referred to is quill-pen-shaped cartilage found in all squids.)

Though Webb urged Dall and Verrill to visit the site and inspect the carcass personally, they did not. Instead, they told Webb to continue his efforts and keep sending them information. However, they ignored the information he sent them. Dall, for instance, kept calling the creature a "cuttlefish" (a cephalopod mollusk related to squids and octopuses, but with ten arms and a hard internal shell).

Webb sent Verrill samples of the creature's remains on February 23. That very day the zoologist declared in letters to both *Science* and the *New York Herald* that the carcass was probably from the "upper part of the head and nose of a sperm whale." Professor Frederic Augustus Lucas of the National Museum examined other samples and decided that they were whale "blubber, nothing more nor less." He also criticized Webb's "imaginative eye" and lack of training. Other cephalopod experts seemed to accept Lucas's explanation. While Webb strongly protested in print, his letters went unanswered. Eventually, the remains of *Octopus giganteus Verrill* rotted away, and the event was forgotten for the next six decades.

Return of the Monster

In 1957 Forrest G. Wood, Jr., curator of the research laboratories of Marineland, Florida, came upon a yellowed newspaper clipping about the Anastasia Island creature. Though an expert on octopuses, Wood had never heard of it.

Fascinated, he launched an investigation that eventually revealed that the Smithsonian Institution still had samples of the animal. They were examined by University of Florida octopus expert Joseph F. Gennaro, Jr., who concluded: "The evidence appears unmistakable that the St. Augustine sea monster was in fact an octopus."

But when Wood and Gennaro published their findings in three articles in the March 1971 issue of *Natural History,* the cause of marine biology was hardly advanced. The editors of the magazine included so many odd and silly comments along with the articles that some readers thought the whole thing a hoax. Wood and Gennaro found out that this was done on purpose—and Wood angrily wrote a letter of complaint to *Natural History,* which it refused to publish. To make matters worse, the *Ocean Citation Journal Index,* which prints briefs of journal articles, stated that the men had concluded that the animal was a *giant squid.* Wood and Gennaro later found out that this misstatement, too, was no accident.

In the mid-1980s, Roy P. Mackal, a University of Chicago biologist, conducted another study of the samples. He found that they were connective tissue and "not blubber." He stated, "I interpret these results as consistent with, and supportive of, Webb and Verrill's identification of the carcass as that of a gigantic cephalopod, probably an octopus, not referable to any known species."

CEPHALOPODS

Cephalopods make up the most highly developed class of mollusks and include squids, octopuses, cuttlefish, and nautiluses. They are free-swimming flesh-eaters that have many muscular, sucker-covered arms or tentacles around the front of the head, which are used to capture prey. A muscular tube or siphon under the head blows water out and allows the animals to move by a kind of jet propulsion. The nautilus has an outside shell. In the squid and cuttlefish, the shell has become smaller and is found inside the body. The octopus has no shell at all. Cephalopods have large heads and eyes, and most have a bag of inky fluid that they can release for defense or camouflage.

Mystery of the Globsters

A carcass discovered in August 1960 on a northwestern beach on the Australian island of Tasmania may have been an animal like the one found in Florida. But here again the investigation was handled badly.

A French
engraving of
unknown date
depicting a
giant octopus
attacking a ship.

Word of the find, made by a rancher and two cowboys working for him, did not reach Hobart, the capital of Tasmania, until months later. First an air search was needed to locate the carcass. Then a four-man scientific team led by zoologist Bruce Mollison of the Commonwealth Scientific and Industrial Research Organization (CSIRO) traveled to the barren site in early March 1962. After Mollison examined the carcass he reported, "One is always seeking some explanation, and you try to add up everything, but this does not add up yet."

The carcass was very odd. Without eyes, head, or bones, it had skin that looked "creamy" and felt "rubbery." It was also "hairy."

Over the next week and a half, the Tasmanian "globster" (a descriptive word created by zoologist Ivan T. Sanderson) appeared in headlines all over the world, and the Australian government was flooded with questions. Under heavy pressure for answers, the government flew a team of zoologists from Hobart to the site for what was supposed to be a complete investigation. But the group returned the next day.

The official report stated that because of the length of time that had passed between the beaching of the animal and its examination, "it is not possible to specifically identify it from our investigations so far." Still, the scientists felt that the remains were "a decomposing portion of a large marine animal" and did not resemble whale blubber. But oddly, the same day he received this report, Senator John Gorton, minister for the Commonwealth of Australia, told the press that "your monster is a large lump of decomposing blubber, probably torn off a whale."

This conclusion disturbed Mollison, who stated that the samples he had taken "could not be identified." And University of Tasmania zoologist A. M. Clark declared that "it was clearly not a whale" (suspecting, instead, a giant ray). Also angered by Gorton's comment was cowboy Jack Boote, one of the animal's discoverers. He suspected the government was trying to cover up the fact that it had acted too slowly on the matter. "They had to say it was nothing new to cover up the fact they hadn't done anything about it before," Boote insisted. "The thing I saw was not a whale or any part of a whale."

None of the laboratory reports, either supporting or denying the whale identification, were ever published. The affair ended in confusion and neglect.

More Globsters

Perhaps these issues could have been settled in 1970 when another globster washed up onto a beach in the same area of northwestern Tasmania. It was found—believe it or not—by the same ranch owner who had come upon the first one: Ben Fenton. Remembering the trouble he had gone through ten years earlier, Fenton was not pleased with his discovery. He told a reporter for the local newspaper, "Be careful you don't quote me as saying it is a monster. I don't know what it is, and I'm making no guesses—not after the last lot." But this time not a single scientist came to investigate.

In March 1965 a globster had appeared on Muriwai Beach on the eastern shore of New Zealand's North Island. It was 30 feet long, eight feet high, and "hairy," according to press accounts. Auckland University zoologist J. E. Morton was quoted as saying, "I can't think of anything it resembles." Another globster washed up on a Mangrove Bay beach in Bermuda in May 1988. Samples were taken of the creature, but results from laboratory tests have yet to be published.

Looking over the cases, J. Richard Greenwell of the International Society of Cryptozoology noted that the "descriptions—and photos—are similar in all cases. All the carcasses were described as tough and hard to cut, usually odorless, and very 'stringy,' which is often called 'hairy.' And, curiously, all seem to be more or less unidentifiable by experts."

Until more is known, it is not certain that the globsters and the St. Augustine creature were the same kind of animal. Nor is it certain that either animal was a giant octopus. Still, this identification remains a possibility.

> "At its thickest the tentacle was as big as a muscular man's upper arm.... It had bumps along it, and one of these hooked on to the edge of the boat."

Sightings

If giant octopuses are real, they would not often be seen, simply because octopuses are bottom-dwelling animals. Still, sightings do occur from time to time. Bahamian fishermen speak of seeing "giant scuttles," and cephalopod expert Forrest Wood, for one, finds their testimony believable.

In late December 1989, press accounts described a frightening Christmas Eve occurrence off Manticao in the southern Philippines. A group of people in a boat carrying an infant's body that was to be buried on a nearby island were startled to see an octopus tentacle flop over the side of the craft. "At its thickest it was as big as a muscular man's upper arm," boat owner Eleuterio Sarino said. "It had bumps along it, and one of these hooked on to the edge of the boat." Another passenger, Jerry Alvarez, said, "I saw other huge tentacles under the water and, though the light was poor even when I used my torch [flashlight], I'm convinced I saw a head down there with big eyes." The tentacles, he claimed, were eight feet long.

The boat began to rock from side to side and then overturned. The passengers waded safely to shore 200 yards away.

In recent years, marine biologists have turned their attention to the extraordinary and sometimes unknown animals that live deep in the world's oceans. Perhaps now more information about this extraordinary creature, if it truly exists, will surface.

Sources:

Gennaro, Joseph F., Jr., "The Creature Revealed," *Natural History* 80,3, March 1971, pp. 24, 84.

Mackal, Roy P., *Searching for Hidden Animals,* Garden City, New York: Doubleday and Company, 1980.

Wood, F. G., "In Which Bahamian Fishermen Recount Their Adventures with the Beast," *Natural History* 80,3, March 1971, pp. 84, 86-87.

Wood, F. G., "Stupefying Colossus of the Deep," *Natural History* 80,3, March 1971, pp. 14, 16, 18, 20-24.

Yoon, Carol Kaesuk, "In Dark Seas, Biologists Sight a Riot of Life," *New York Times,* June 2, 1992.

GIANT SQUID

On November 30, 1861, as they sailed in the Canary Islands, the crew of the French gunboat *Alecton* came upon a giant sea monster. When the sailors tried to capture it, the creature swam off. Bullets and cannonfire failed to stop it. After a long chase the ship got close enough so that a harpoon could be hurled into the creature's flesh. Then a noose was put around its body, but the rope slipped until it reached the back fins. As the crew tried to lift it into the ship, the monster's body broke free. All but a small part of the tail slipped back into the water.

On landing at the island city of Tenerife, the *Alecton*'s commander contacted the French consul there and showed him the tail specimen. He also wrote up an official report, which was read at the December 30 meeting of the French Academy of Sciences. It was not well received; speaking for academy members, Arthur Mangin stated that no "wise" person—"especially the man of science"—would report such an extraordinary creature, whose very existence went against the laws of nature.

In other words, the ship's crew had to be lying or imagining the incident. In the *Alecton* case, though, witnesses had the unhappy experience of coming upon a strange but real animal a few years before its existence would be officially recognized. The crew had seen a giant squid—and a fairly small example of one. It was about 24 feet long from the tip of its tail to the end of its arms. Much larger squids are known to exist, and ones of *gigantic* size are suspected.

In 1861 the crew of the French steamer *Alecton* tried unsuccessfully to capture a giant squid.

Early Descriptions

In the eighteenth century, Bishop Erik Pontoppidan wrote about the Kraken, a legendary monster of the north seas, in his important zoological book *The Natural History of Norway* (1752–53). Though Pontoppidan did some exaggerating (he said the creature measured "about an English mile and a half" across and its arms could pull the "largest man-of-war [warship] ... down to the bottom"), he described the giant squid fairly accurately.

Other early descriptions of the giant squid were usually treated as mere imagination and folklore. Thus a very early account of a stranding and killing of a giant squid in Ireland's Dingle Bay in October 1673 attracted little attention when it was published. But now, the fantastic details with the giant squid in mind makes a great deal of sense. The report spoke of a beast with "two heads and Ten horns, and upon ... the said Horns about 800 Buttons, ... and in each of them a set of Teeth." The creature was 19 feet long, with a body "bigger than a Horse, ... and two very large Eyes." Squids, of course, have only one head, but the "little head" is the siphon through which water is pumped to propel the animal. The "horns" are the arms or tentacles, and the "buttons" are the toothed suckers on them.

The first scientist to undertake a complete study of the Kraken was nineteenth-century Danish zoologist Johan Japetus Steenstrup. He uncovered records of what sounded like giant squid strandings as early as 1639 (on Iceland's coast). He also collected pieces of specimens and gave a lecture on the subject to the Society of Scandinavian Naturalists in 1847. But his talk stirred little interest. Six years later Steenstrup obtained the pharynx and beak of a specimen from fishermen who cut up the rest, as they usually did, for bait (to them the Kraken had always been both real and useful). The zoologist published a description of the animal and gave it its scientific name, *Architeuthis,* in 1857.

Steenstrup's work continued to be ignored. The collective testimony of the *Alecton* crew did not help his case in the least. Zoology textbooks paid no attention to Steenstrup's new animal—until the 1870s, when a series of strandings on Canada's Newfoundland and Labrador shores brought some open-minded scientists, including *American Naturalist* editor A. S. Packard, to investigate.

In October 1873, a fisherman named Theophile Piccot and his son chopped off a tentacle of a giant squid they had come upon in the waters off Great Bell Island near Saint John's, Newfoundland. Piccot told Geological Commission of Canada investigator Alexander Murray that 10 feet of the tentacle had been left on the body; their piece measured 25 feet. Piccot claimed the animal was immense: roughly 60 feet long and from five to ten feet across.

Giant and Beyond

In the decades since, with the mystery of the giant squid effectively solved, other questions have emerged. For instance, what do the creatures eat, how do they live, and how do they reproduce? No live specimen has ever been captured for lengthy scientific observation. But the most pressing question of all is: how giant can a giant squid get?

The fact that the giant squid's primary enemy is the sperm whale might offer a clue. (The male sperm whale can reach 70 feet in length.) A rare witnessed battle between these two giants of the ocean was said to occur late on an evening in 1875 at the entrance to the Straits of Malacca (connecting the Indian Ocean with the South China Sea). Frank T. Bullen's *The Cruise of the Cachalot* (1924) described the event from an eyewitness's point of view:

> There was a violent commotion in the sea.... Getting the night-glasses out of the cabin scuttle, ... [I saw a] very large sperm whale was locked in deadly conflict with a cuttle-fish, or

Scene from the 1954 Walt Disney film *20,000 Leagues under the Sea.*

A giant squid beached in 1980 on Plum Island, Massachusetts.

squid, almost as large as himself, whose interminable tentacles seemed to enlace the whole of his great body. The head of the whale especially seemed a perfect network of writhing arms—naturally, I suppose, for it appeared as if the whale had the tail part of the mollusc in his jaws, and, in a business-like, methodical way, was sawing through it. By the side of the black columnar head of the whale appeared the head of the great squid, as awful an object as one could well imagine even in a fevered dream.... The eyes were very remarkable from their size and blackness.... They were, at least, a foot in diameter, and, seen under such conditions, looked decidedly eerie and hobgoblinlike.

Even without such remarkable eyewitness testimony, we know that squid–whale battles take place for two reasons: squid remains found in whale stomachs and vomit and sucker scars on whales. Both of these can help us guess how large squids may become.

The largest squid specimen documented by science was found on a New Zealand beach in 1880 and measured about 65 feet. Two scientists who investigated reported that much of the length (30 to 36 feet) "consisted of the tentacles." But the scientists also noted that "dead squid are notably elastic and easily stretched," and that fact kept their measurements from being entirely reliable. Still, the animal was huge. Other eyewitness accounts have reported 80- to 90-foot specimens.

Though direct sightings of giant squids are rare and poorly documented, many whalers have reported seeing amazing materials vomited up by sperm whales as they are dying. Bullen saw a "massive fragment of cuttle-fish—tentacle or arm—as thick as a stout man's body, and six or seven sucking-discs or *acetabula* on it. These were about as large as a saucer, and on their inner edge were thickly set with hooks or claws all around the rim, sharp as needles, and almost the shape and size of a tiger's."

Sucker Scars

Before scientists recognized that the Kraken was real, the strange round marks that they found on sperm whales puzzled them. Eventually they learned that they were sucker scars, made by giant squids locked in brave but losing battles with whales determined to eat them. Scars have been found to measure as much as 18 inches across.

Some teuthologists (zoologists who study cephalopods: squids, cuttlefish, and octopuses) argue that sucker scars are unreliable for judging the size of squids; in the words of Clyde F. E. Roper and Kenneth J. Boss, "a scar grows as a whale grows." But other zoological writers disagree. Bernard Heuvelmans, the founder of cryptozoology, noted that "scars are rare on female whales" and that "a baby whale would be kept well away from such huge brutes, and, if attacked, would hardly survive." In other words, giant squids are most likely to leave their marks on fully grown adult male sperm whales.

At any rate, there is no shortage of testimony concerning extraordinary squid remains in whale bellies. One ship captain noted an arm or tentacle 45 feet long and two and one-half feet thick, and others have reported those in the 25- to 35-foot range.

Giant squids, which are rarely seen, spend most of their lives in moderately deep to very deep waters. (The strandings seem to occur when a sick squid dies and rises to the surface, washing to shore.) A complete scientific survey of the ocean depths has only now just begun, and it is thought that merely one-tenth of one percent of it has been studied so far. Some of the scientists involved in this research hope especially to see a giant squid and even larger and stranger creatures.

> One ship captain noted an arm or tentacle 45 feet long and 2 1/2 feet thick, and others have reported those in the 25- to 35-foot range.

Sources:

Heuvelmans, Bernard, *In the Wake of the Sea-Serpents,* New York: Hill and Wang, 1968.
Ley, Willy, *Exotic Zoology,* New York: The Viking Press, 1959.
Yoon, Carol Kaesuk, "In Dark Seas, Biologists Sight a Riot of Life," *New York Times,* June 2, 1992.

SEA SERPENTS

The American ship *Silas Richards* was sailing off St. George's Bank south of Nova Scotia, Canada, at 6:30 P.M. on June 16, 1826, when its captain, Henry Holdredge, and a passenger, Englishman William Warburton, saw the oddest sight: a huge, many-humped snakelike creature moving slowly toward the boat. Warburton raced below deck to tell the other passengers, but only a handful of them were interested. "The remainder refused to come up," Warburton recalled, "saying there had been too many hoaxes of that kind already."

Over several centuries, the mystery of the sea serpent has been the subject of heated debate. Despite credible accounts from reliable witnesses going back hundreds of years, "the great unknown," as the creature was once called, has often been blamed on mistakes, lies, or wild imaginings. By the time of the *Silas Richards* sighting, the sea serpent had become—in the words of cryptozoology pioneer Bernard Heuvelmans—the "very symbol of a hoax."

Some serpents were reported to have large foreheads, some had pointed snouts, and some had flat snouts "like that of a cow or horse, with large nostrils, and several stiff hairs standing out on each side like whiskers."

Early History

Though sea serpents have long appeared in myths and legends, the first description of one as a real animal appeared in a 1555 work by Olaus Magnus, a Catholic archbishop of Uppsala, Sweden. He wrote that sailors off the coast of Norway had often seen a "serpent ... of vast magnitude, namely 200 feet long, and moreover 20 feet thick." A dangerous beast, it lived in caves along the shore and fed on both land and ocean creatures, including—from time to time—the unlucky seaman!

Historians who followed Magnus also noted that "serpents" were seen regularly in the North Sea (the part of the Atlantic Ocean between Europe and Great Britain), though not everyone considered them dangerous. In 1666 Adam Olschlager wrote of a sighting of a "large serpent, which seen from afar, had the thickness of a wine barrel, and 25 windings. These serpents are said to appear on the surface of the water only in calm weather and at certain times."

In 1734 a Protestant priest, Hans Egede, saw a "monster" about 100 feet long rise from the water off the coast of Greenland. He recorded the experience in a book published in 1741. A little more than a decade later, Bishop Erik Pontoppidan wrote *The Natural History of Norway* (1752-53), an important book that addressed sea serpents, **merfolk** (mermaids and mermen), and the Kraken (**giant squid**).

Because of the testimonies of reliable people, Pontoppidan believed that all of these doubted creatures were real. He also felt that sea serpents involved more than one type of animal, for descriptions of them included different details. Some serpents were reported to have large foreheads, some had pointed snouts, and some had flat snouts "like that of a cow or horse, with large nostrils, and several stiff hairs standing out on each side like whiskers." Over the next two centuries Pontoppidan's writings on sea serpents would be referenced again and again in discussions of the subject.

In the Americas

In *An Account of Two Voyages to New England* (1674), John Josselyn recalled a 1639 conversation with members of the Massachusetts colony. "They told me of a sea-serpent or snake, that lay coiled upon a rock at Cape Ann," he wrote. This is the first known printed account of an American sea serpent. In the next century and a half, thousands of residents of New England and Canada's coastal provinces would observe similar creatures.

One such account was from Captain George Little of the frigate *Boston.* In May 1780, while in Broad Bay off the Maine coast at sunrise, he "discovered a huge Serpent, or monster, coming down the Bay, on the surface of the water." Deciding to chase the creature, Little and several crew members boarded an armed cutter, and when it got to within 100 feet of the beast, he gave the order to fire. But at the same time the serpent dove below the surface of the water. "He was not less than from 45 to 50 feet in length," Little related. "The largest diameter of his

Sea serpent seen frequently off Gloucester, Massachusetts, in the early nineteenth century.

body, I should judge, was 15 inches; his head nearly the size of that of a man, which he carried four or five feet above the water. He wore every appearance of a common black snake."

A year earlier the crew of the American gunship *Protector* had also had an extraordinary experience off the Maine coast, in Penobscot Bay. One of the witnesses was 18-year-old ensign Edward Preble, who would go on to become a notable figure in U.S. naval history. When *Last of the Mohicans* author James Fenimore Cooper wrote Preble's biography, he included an account of the strange event:

> The day was clear and calm, when a large serpent was discovered outside the ship. The animal was lying on the water quite motionless. After inspecting it with the glasses [binoculars] for some time, Capt. Williams ordered Preble to man and arm a large boat, and endeavor to destroy the creature; or at least to go as near to it as he could.

> Preble shoved off, and pulled directly toward the monster. As the boat neared it, the serpent raised its head about ten feet above the surface of the water, looking about it. It then began to move slowly away from the boat. Preble pushed on, his men pulling [their oars] with all their force, and the animal being at no great distance, the swivel was discharged loaded with bullets. The discharge produced no other effect than to quicken the speed of the monster, which soon ran the boat out of sight.

There would be other sightings in the following decades. But New England's sea serpent would not become a matter of worldwide interest until the second decade of the nineteenth century. Over a period of several years, from Boston up to Cape Ann at the northeastern tip of Massachusetts, many witnesses on both ship and shore would see the animal.

Witness Solomon Allen III, for instance, saw the creature on August 12, 13, and 14, 1817, in Gloucester harbor. Judging the sea serpent "to be between eighty and ninety feet in length," Allen thought its head looked snakelike, but it was "nearly as large as the head of a horse." He also noted that when the creature "moved on the surface of the water, his motion was slow, at times playing about in circles, and sometimes moving nearly straight forward. When he disappeared, he sunk apparently down."

On June 6, 1819, Hawkins Wheeler also got a clear view of the sea serpent. "The creature was entirely black; the head, which perfectly resembled a snake's, was elevated from four to seven feet above the surface of the water, and his back appeared to be composed of bunches or humps, ... I think I saw as many as ten or twelve." Wheeler thought the humps were caused by the wavy motion of the animal; he guessed its length at 50 feet. Also that year, on August 14, Samuel Cabot noticed the serpent's "eight or ten regular bunches" or humps as well. Guessing the creature's length at 80 feet, Cabot similarly noted that it carried its "serpent shaped" head above the water.

A Sea Serpent Investigation

As the result of these sightings, the Linnean Society of New England met in Boston on August 19, 1817, to lay out a plan of investigation. The group selected three men—a judge, a physician, and a naturalist—to interview witnesses and get sworn statements from them. In the meantime, sea serpent sightings continued. The testimony that the men collected, and further accounts from later witnesses led to this general description of the creature: It was huge, snakelike, dark on the top and lighter on its underside, moving with vertical waves or undulations.

The animal, however, could not have been a serpent. Snakes and other reptiles move from side to side, not up and down. Regardless, the society investigators concluded that the animal was a huge reptile that was remaining close to shore because it had laid its eggs there. Repeated searches revealed no such eggs. But when a farmer killed a three-foot black snake in a field just off Cape Ann, he noticed it had a series of bumps along its back—just as the sea serpent was reported to have.

> It was huge, snake-like, dark on the top and lighter on its underside, moving with vertical waves or undulations.

The Linnean Society foolishly agreed with the farmer's suggestion that this was a recently hatched baby sea serpent. Afterward, another scientist, Alexandre Lesuerur, demonstrated that the specimen was no more than a deformed specimen of the common black snake. Though Lesuerur did not intend to make the entire sea serpent investigation seem senseless, his discovery was hailed by nonbelievers and doubting journalists, and the entire affair ended in disaster for hopeful investigators.

The Great Unknown

No amount of laughter, however, could stop the sightings, which kept coming in from all over the world. But fear of ridicule did stop some people from reporting them. When the great American statesman Daniel Webster saw a sea serpent while on a fishing trip off the Massachusetts coast, he pleaded with his companion, "For God's sake never say a word about this to anyone, for if it should be known that I have seen the sea serpent, I should never hear the last of it."

Despite all attempts to explain away sightings of sea serpents, there were still some scientists who supported their existence. While most sea serpent accounts were reported in newspapers, a few made their way into scientific journals. But for every scientist who felt that his "evidence" demonstrated that sea serpents were real, there were ten times as many who had "proof" that the creatures were not. Finally, in 1847 *Zoologist* editor Edward Newman made a bold move by opening the pages of his journal to a fair-minded discussion of the subject. In addition, he scolded nonbelievers for ignoring "fact and observation" simply because the sea serpent "ought not to be."

Not Just a Fish Story

Perhaps the most famous sea serpent report of all time took place a year later. It occurred late in the afternoon of August 6, 1848, and the witnesses were the captain and crew of the frigate *Daedalus,* on their way back to England from Africa's Cape of Good Hope. Soon after the ship's arrival at Plymouth on October 4, several newspapers reported rumors that the captain and crew had experienced a spectacular 20-minute sea serpent encounter. Navy officials asked Captain Peter M'Quhae to either deny or describe the event.

On October 11 M'Quhae told Admiral Sir W. H. Gage about the incident in a letter, which was reprinted in the *London Times.* He, along

THE 'DÆDALUS' SEA-SERPENT

with other officers and crew members, had spotted "an enormous serpent, with head and shoulders kept about four feet ... above the surface of the sea.... There was at the very least 60 feet of the animal." The witnesses were unable to tell how the creature moved itself through the water, even though it came so close to their boat that, in M'Quhae's words, "had it been a man of my acquaintance, I should easily have recognized his features with the naked eye." Moving at a speed of about 12 to 15 miles per hour, "it was never," according to the captain, "during the 20 minutes that it continued in sight of our glasses, once below the surface of the water." Fifteen or 16 inches around, the serpent had the head of a snake. "It had no fins, but something like the mane of a horse, or rather a bunch of seaweed, washed about its back."

Soon afterward the *Zoologist* published the private notes of another witness, Lieutenant Edgar Drummond. His account matched M'Quhae's except for one detail: he thought the "mane" along the creature's back looked more like a fin. Ten years later another officer recalled the event in a letter to the *Times*. To him the animal appeared more like a lizard than a snake, "as its movement was steady and uniform, as if propelled by fins, not by any undulatory power."

These accounts caused an uproar. Sea serpent disbelievers were especially alarmed that respectable, experienced British officers would report such a sight. They scrambled to provide other explanations. One doubter felt that M'Quhae and his crew had mistaken a patch of seaweed for a creature. Another, the great anatomy scientist Sir Richard Owens, suggested that the sailors had come upon a giant seal and that their excitement and imaginations had supplied the "unseal-like" details.

Writing in the *Times,* M'Quhae boldly defended himself and his crew against the famous professor Owens, who happened to be the navy's consultant on sea serpents (and who said that they were less likely to exist than ghosts). The captain stood by every detail of his story and—in the opinion of many—won the argument. Still, scientific support for the sea serpent seemed to dwindle, despite the great many reports from reliable men and women.

The point was also raised that if sea serpents existed, why did they never get stranded on beaches, their dead bodies settling the mystery once and for all? This was a good question but one full of problems. For when unusual remains *were found,* scientists sometimes refused to examine them, so strong were their negative feelings about sea serpents. On the other hand, when examinations were carried out, the bodies usually ended up being those of known sea animals. This further weakened the case for the sea serpent's existence.

In 1892 A. C. Oudemans revived the fading mystery with the publication of his important book *The Great Sea Serpent,* the most careful and complete study of the subject yet to appear. In 591 pages the respected Dutch zoologist examined 187 cases, concluding that all sea serpent sightings were of a single species of animal, a gigantic long-necked seal.

In 1933 reports of strange animals in a Scottish lake caused a sensation; in fact, the **Loch Ness monster** would capture worldwide attention. For a period of time, the Ness story reminded scientists and others of the still-unsolved mystery of the sea serpent. Oudemans, for one, believed that one of the Ness animals would soon be caught or killed, and this would finally reveal the identity of the sea serpent as well.

Cadborosaurus

There was also a flurry of reports of a sea serpent off the coast of British Columbia in 1933. Sightings had occurred there in the past, going back to at least 1897, but the Loch Ness uproar made water monsters popular again. Soon the Canadian animal was given the name Cadborosaurus, which combined its home, Cadboro Bay, with the Greek word for lizard, *sauros*. Cadborosaurus soon became "Caddy."

The first widely reported sighting of Caddy took place on October 8, 1933, and involved a very reliable witness: Major W. H. Langley, a legal counselor and member of the British Columbia government. Sailing his sloop past Chatham Island early in the afternoon, he spotted a greenish-brown serpent with a sawlike body "every bit as big as a whale." He guessed it was 80 feet long.

In 1937, according to an account related to investigators years later, whalers killed a sperm wale off the Queen Charlotte Islands in northern British Columbia. When they cut open its stomach, they found the half-digested remains of a 10-foot-long snakey creature with a horselike head and humped back. They threw the remarkable specimen back into the ocean.

Reports like these captured the interest of two scientists, University of British Columbia oceanographer Paul LeBlond and Royal British Columbia Museum marine biologist Edward Bousfield, who over the years investigated a great many of them. By 1992 they had become so convinced that sea serpents were real that they said so in a formal lecture at a meeting of the American Society of Zoologists.

Chessie

Another well-known sea serpent was Chessie, the Chesapeake Bay monster in Maryland. It got its name in 1982, following a number of sightings in the spring and summer. One of these was made by Robert and Karen Frew on May 31, 1982. At 7 P.M., while entertaining guests outside their home overlooking the bay at Love Point on the northern tip of Kent Island, they saw a strange creature 200 feet from shore in calm water only five feet deep.

Robert Frew watched the animal through binoculars for a few minutes before getting his video camera. Then he focused on the creature, which disappeared below the surface of the water several times during the sighting. The closest it came to shore was around 100 feet, though it did come to within 50 feet of some boys who were playing on a pile of underwater rocks. Though the Frews and their friends shouted to alert the boys (their efforts captured on the videotape), the children never heard them—or saw the animal.

The witnesses guessed that the creature was 30 to 35 feet long and about a foot around. While much of it remained under water, each time it reappeared above the surface they could make out a little bit more. They could see humps on the back. The head was shaped like a football, only "a little more round." It was the odd shape of the head, in fact, that made Robert Frew think that the creature was something other than a snake. Familiar with a wide variety of sea life, the Frews were sure they had not mistaken a known animal for a mysterious sea creature.

On August 20 seven scientists from the Smithsonian Institution in Washington, D. C., along with members of the National Aquarium and Maryland's Department of Natural Resources, met to view and discuss the Frew videotape. The Smithsonian's George Zug reported the group's conclusion: "All viewers of the tape came away with a strong impression of an animate [living] object.... We could not identify the object.... These sightings are not isolated phenomena, for they have been reported regularly for the past several years."

Other Types of Sea Serpents

And reports of sea serpents in the world's oceans continue. Since the 1982 founding of the International Society of Cryptozoology (ISC), the membership of which includes a number of notable biologists, serious research into the subject has been made both possible and even somewhat respectable. ISC president Bernard Heuvelmans, in fact, published the most complete volume ever written about the mystery, the enormous *In the Wake of the Sea-Serpents,* in 1968.

In his book Heuvelmans described and examined every known sea serpent account, believable or otherwise, through 1966. There were 587 of them! Of those, he judged 358 to be real observations of unknown animals. Unlike most investigators of sea serpent mysteries, he decided that the sightings did not describe a single species of animal. The differences in details could not just be explained away as mistakes or imagination—they were important. But even though details varied, enough were repeated to create patterns. This led Heuvelmans to believe that several separate, unknown water animals were being describe, among them the following:

Long-necked (48 sightings)

Description: A long neck angled toward the head; hump or humps on the back; no tail; two horns, sometimes described as ears. **Classification:** Almost certainly a pinniped. (A pinniped is a flesh-

Sea monster photographed by Robert Le Serrec, at Stonehaven Bay, Hook Island, Australia, in 1964.

eating water mammal with four flippers for limbs, like a seal or a walrus.) **Range:** Widespread.

Merhorse (37 sightings)

Description: Floating mane; medium to long neck; big eyes; hair or whiskers on face. **Classification**: Probably a pinniped. **Range**: Widespread.

Many-humped (33 sightings)

Description: String of humps on back; slender neck of medium length; small but striking eyes; dark stripes on top of the body, white on underside, white stripes on neck. **Classification:** Cetacean. (A cetacean is a water mammal with a large head, fishlike—nearly hairless—body, and paddle-shaped front limbs, like a whale, dolphin, or porpoise.) **Range:** North Atlantic.

Many-finned (20 sightings)

Description: Triangular fins that look like large peaks; short, slender neck. **Classification**: Cetacean. **Range**: Tropical waters.

Super-Otter (13 sightings)

Description: Slender, medium-length neck and long, tapering tail; several vertical bends in the body. **Classification:** Uncertain, but possibly a surviving form of primitive cetacean. **Range:** North Atlantic (possibly extinct; last known sighting in 1848).

Super-Eel (12 sightings)

Description: Serpent's body; long tapering tail. **Classification**: Fish. **Range**: Widespread.

While Heuvelmans knew he had not solved the problem of the sea serpent with his book, he felt that he had shed a great deal of light on the subject. "To solve the whole complex problem, without being able to examine the remains of the animals in question, we need many more detailed and exact reports," he noted. When conducting his study, he had been forced to leave out many accounts because of lack of detail. Heuvelmans also counted 49 hoaxes and 52 mistaken identifications among the reports he collected.

Like every other twentieth-century investigator, Heuvelmans did not think that sea serpents were actual serpents or reptiles. Mammals were the most frequent candidates for the many cases that he studied. But he agreed that an unknown reptile might have been sighted on rare occasions. His book contained just four reports of what he called a "marine saurian," a huge lizard- or crocodile-shaped creature spotted in tropical waters.

According to Heuvelmans, if such a creature existed, it might be "a surviving thalattosucian, ... a true crocodile of an ancient group, a specifically and exclusively oceanic one, which flourished from the Jurassic to the Cretaceous periods. But it could also be a sur-

---

SEA SERPENT HOAX

One of the most famous sea serpent hoaxes took place a decade after *In the Wake of the Sea-Serpents* appeared. Photographs of the Cornish sea creature Morgawr, from England's Falmouth Bay, were widely published. Supposedly taken by an untraceable "Mary F.," the pictures are believed to be those of professional prankster Tony "Doc" Shiels, who has also taken questionable photos of the Loch Ness monster.

viving monosaur, a sea cousin of the monitors [a dragonlike, mostly tropical, lizard] of today. It would not be surprising if it had survived for so long at sea, since it is well designed to dive deep and remain unseen."

And what answer did Heuvelmans have for the frequently asked question of why sea serpents don't get stranded on beaches? The beasts responsible for such sightings, he wrote, "all belong by nature to the category of animals least likely to be stranded, and quite capable of getting off the shore again, if by misfortune they are." So it appears that such creatures probably die far out to sea.

As time goes on, scientists are learning more and more about the mysterious plants and animals that live deep and unobserved in the world's oceans. The sea serpent's time may be here at last.

Sources:

Costello, Peter, *In Search of Lake Monsters,* New York: Coward, McCann and Geoghegan, 1974.

Gould, Rupert T., *The Case for the Sea Serpent,* London: Philip Allan, 1930.

Heuvelmans, Bernard, *In the Wake of the Sea-Serpents,* New York: Hill and Wang, 1968.

Yoon, Carol Kaesuk, "In Dark Seas, Biologists Sight a Riot of Life," *New York Times,* June 2, 1992.

Freshwater Monsters

- LAKE MONSTERS

- LOCH NESS MONSTERS

- CHAMP

- MORAG

- OGOPOGO

- WHITE RIVER MONSTER

Freshwater Monsters

LAKE MONSTERS

At 7 P.M. on February 22, 1968, farmer Stephen Coyne arrived at the dry bog near Lough Nahooin, Ireland, one of a series of small lakes linked by streams that run through Connemara. With him were his eight-year-old son and the family dog. In the water he noticed a black object, and thinking it was his dog, he whistled for it. But the dog came bounding up from somewhere else. The moment it saw the object in the water, it stopped and stared.

The object proved to be a strange animal with a narrow, polelike head (without visible eyes) and a neck nearly a foot around. It was swimming in many directions, thrusting its head and neck underwater from time to time. Whenever this happened, two humps on its back and—sometimes—a flat tail would come into view. Once the tail was seen near the head, showing that the animal was both long and flexible. Its skin was black, slick, and hairless. The creature appeared to be at least 12 feet long.

One time, bothered by the dog's barking, it swam toward the group, its mouth open. Coyne stepped forward to protect the dog and the creature retreated, continuing its casual, directionless movement through the water. Soon father and son were joined by the five other members of the Coyne family. The animal remained clearly in view, sometimes as close as five or six yards. It was still there when darkness fell and the Coynes decided to go home.

The Coyne family's experience is one of the most credible in a long history of lake monster sightings. These have ranged from the clearly phony to the seemingly real. In the Coyne's case, a team of experienced

LAKE MONSTERS AND FOLKLORE

Since ancient times reports of giant freshwater "monsters" of various names and descriptions—great serpents, dragons, water horses, and many others—have been abundant in fairy tales, myths, and folklore. But to most researchers, the ancient accounts are generally not useful for zoological discovery. Our ancestors lived in a world in which magic and the supernatural served as explanations for many puzzles. Although there may have been sightings of the same creatures in ancient and contemporary times, to interpret medieval magical traditions of evil dragons and sea serpents as if they were literally true—in modern terms—is probably an error. If unusual and unrecognized freshwater animals do exist today, the sightings of them that have been reported over the last two centuries are, to most scientists, the most effective way to discover their traits.

cryptozoological investigators, including University of Chicago biologist Roy P. Mackal, interviewed the adult and child witnesses soon after the sighting. They found their testimonies solid.

A few months later, as the investigators were trying to snare the creature by dragging the tiny lake, they met Thomas Connely, a local man who had seen the same or a similar creature in September as it plunged into the water from the bank. They also heard reports from other lakes in this remote area of western Ireland.

In a case like this, mistaken identification or runaway imagination seems as unlikely as the creature the Coynes claimed to have seen. Sightings of lake monsters are frequent in the modern world. Some of the reports (often well detailed and with accompanying photographs) are known or suspected to be fake. Some of the reports can be reasonably explained. And some reports are just plain mysteries.

Water Horses

Some contemporary writers on lake monsters have linked modern reports—especially those from the monster-haunted lochs and loughs (Gaelic words for lakes) of Scotland and Ireland—to folklore about "water horses." They argue that in these legendary tales lies hidden evidence that unusual lake creatures really do exist. The link is a shaky one, though. The only two things that lake monster reports and water horse legends have in common are their freshwater homes and that many lake monsters are reported to have heads that look like that of a horse. Beyond this connection, the water horse (known as the "kelpie" in the Scottish Highlands) is a separate creature altogether.

The water horse is believed by many to be a dangerous shape-changer that can appear either as a shaggy man who leaps out of the dark onto the back of a lone traveler and frightens or crushes him to death or as a young horse, which after tricking an unknowing soul onto its back plunges to the bottom of the nearest lake, killing its rider.

Though water horses are discussed at length in folklore writings, it is almost impossible to find "sightings" of them; rumors and folktales supply the details about the creatures' appearance and habits. Yet one rare "sighting" was reported by Mary Falconer of Achlyness, West Sutherland, Scotland, one afternoon in the summer of 1938. Walking with a companion near Loch Garget Beag, she noticed a herd of 13 ponies grazing near the water. Because she was carrying a heavy sack of venison, Falconer thought she would borrow one of the horses for the rest of her trip to Rhiconich. The idea came to her because she spotted among the animals a white horse that looked exactly like her neighbor's.

But as she approached the animal, she found that it was too big to be her friend's horse. When she saw that it had water weeds tangled in its mane, she knew at once that it was a water horse. At that moment, it and the rest of the herd raced for the lake and disappeared below the surface! According to folklorist R. Macdonald Robertson, Falconer's companion supported every detail of her story.

> ## PLESIOSAURS
>
> Plesiosaurs were a suborder of prehistoric reptiles that dominated the seas during the Cretaceous period (136 to 65 million years ago). Their bodies were short, broad, and flat. They had short, pointed tails. Their small heads were supported by long, slender necks, ideal for darting into the water to catch fish. Plesiosaurs swam with a rowing movement, using their four powerful, diamond-shaped flippers like paddles. They were often quite large, measuring up to 40 feet in length.

Appearance

The twentieth-century image of a lake monster is that of a long-necked, plesiosaurlike animal. But before our modern age, with its scientific instruments and photographic evidence, freshwater monsters were usually described as great serpents. These serpents did not always remain in the water and were often dangerous.

In 1636, for example, Norwegian churchman Nicolas Gramius reported that a great serpent that had lived in the Mjos and Branz rivers made its way to the sea during a flood. "From the shores of the ... river, he crossed fields," Gramius related. "People saw him moving like a long ship's mast, overturning all that he met on his path, even trees and huts."

In more recent times (as in sightings reported in central Wisconsin in the 1890s), the lake monster has been described as 10 to 20 feet long, snake-shaped, and moving with an undulating or wavy motion. Snakes and other reptiles move from side to side, not up and down, but this wavy movement has nonetheless been described in many reports. Such accounts, while sketchy, survive in local newspapers. Mackal and

Representation of a plesiosaur.

many other trained cryptozoologists feel that this description of a lake monster is consistent with descriptions of the *zeuglodon,* a primitive, snakelike whale believed to no longer exist.

The Freshwater Dragon

The freshwater "dragon," on the other hand, surfaced as recently as October 18, 1946, in the Clearwater River near Rocky Mountain House in Alberta, Canada. Farmer Robert Forbes claimed to have seen a huge, horned, scaly-skinned monster with fiery eyes and long, flashing teeth dart its head out of the water long enough to swallow–whole–a calf that happened to be grazing on the bank.

Creature-Haunted Lakes

Lake monsters are reported worldwide, but are poorly documented. A list of the earth's "creature-haunted" lakes–about 300 of them–appeared in the spring 1979 issue of *Pursuit* magazine. Because many of these lakes got their reputations from questionable nineteenth-century newspaper stories and frontier tall tales, the number is not a dependable one.

Outside North America and the British Isles, most serious investigation has focused on Scandinavian lakes, especially in Norway. While these efforts have brought no solid results, some of the eyewitness reports there have been believable and impressive. Most evidence based on instrument findings has come from Loch Ness, where investigations began in the 1930s and continue today. Films, photographs, and sonar (sound wave) trackings have drawn a great deal of attention to "Nessie" (the **Loch Ness monster**) and strongly suggest that *something* unusual is going on in Scotland's most famous lake.

> **MIGRATING MONSTERS**
>
> Norwegians once believed that monsters grew in lakes until they were too big to live there any longer; then they migrated to the sea. It is not entirely impossible that these monsters were large eels, which have been known to migrate as much as 20 miles overland.

Sources:

Bord, Janet, and Colin Bord, *Alien Animals,* Harrisburg, Pennsylvania: Stackpole Books, 1981.
Coleman, Loren, *Mysterious America,* Boston: Faber and Faber, 1983.
Heuvelmans, Bernard, *In the Wake of the Sea-Serpents,* New York: Hill and Wang, 1968.
Holiday, F. W., *The Dragon and the Disc: An Investigation into the Totally Fantastic,* New York: W. W. Norton and Company, 1973.

LOCH NESS MONSTERS

The first report of Scotland's lake monster "Nessie" is traced by believers back to A.D. 565. A man swimming in the river Ness, at the north end of the loch (lake), was reportedly killed by the beast. St.

Columba, so the story goes, came upon some men carrying the body and was told about the unusual death. He sent a companion into the river, which attracted the attention of the creature; it rose up and moved threateningly toward the swimmer. As the others looked on in terror, Columba formed the sign of the cross and commanded the monster to depart in the name of God. According to a Latin text compiled by St. Adamnan a century later, the "beast, on hearing this voice of the saint, was terrified and fled backwards more rapidly than he came."

The account did not describe what the "beast" looked like, nor was its fierce behavior typical of the loch's modern monster. Still, those devoted to the Ness mystery considered the story important. Other unclear references to large animals in the loch appeared in documents over the centuries, with equally weak links to contemporary Nessie reports. Some writers on lake monsters, for instance, tried to connect widespread European folktales about "water horses"—known in Scotland as "kelpies"—with modern Ness sightings (also see **Lake Monsters**). The only feature that the supernatural kelpies and some modern Ness monsters share is their horselike heads.

> ### LOCH NESS
>
> Loch Ness is a freshwater lake in the Highland region of northwest Scotland. It is over 20 miles long, 1½ miles wide, and—in places—1,000 feet deep. It is also thought to be the home of the world's best-known lake monsters.

Thousands of Sightings

The Loch Ness monster became a worldwide sensation in the 1930s, but there had been many sightings before that time. During the 1930s, residents of the loch area and others hearing the reports came forward with their own sightings from earlier in the century or before. In 1934, for instance, D. Mackenzie wrote to Rupert T. Gould, author of *The Loch Ness Monster and Others* (1934), the first book on the Ness mystery. In the letter Mackenzie recalled a sighting in 1871 or 1872 when, on a sunny October day, he saw what looked "rather like an upturned boat ... wriggling and churning up the water." On October 20, 1933, *The Scotsman* published a letter from the duke of Portland, who remembered that in 1895, while involved with salmon fishing in the area, he had heard many mentions of "a horrible great beastie ... which appeared in Loch Ness." And two groups of witnesses recalled seeing a large elephant-gray animal with a small head at the end of a long neck as it "waddled" from land into the water back in 1879 and 1880.

Such descriptive terms as upturned boat, elephant-gray color, small head, and long neck would commonly appear in later Nessie

Rupert T. Gould

(1890-1948)

Though largely forgotten today, Rupert Thomas Gould was a pioneering writer on mysterious phenomena and a popular public figure in England. He wrote the first book on the Loch Ness monster and would go on to write two books about sea serpents and two best-selling books, *Oddities* (1928) and *Enigmas* (1929), which covered a wide range of anomalies.

Born in England and educated at Dartmouth College, Gould worked for the Royal Navy for many years. In the 1930s he hosted a BBC radio program for children called *The Stargazer*. His show, on which he discussed the many mysteries of science and history that so fascinated him, was immensely successful.

Though Gould's life and career overlapped with American anomalist Charles Fort's, the two men could hardly have been more different in approach (also see entry: Falls from the Sky). Fort used his data on a vast range of anomalies to create a sometimes outrageous worldview that mocked the pretensions of scientists and scholars. Gould, on the other hand, chose his subjects carefully, learned everything he could about them, and proceeded scientifically and logically. While Fort came up with revolutionary ideas, Gould was the more reliable and accurate anomalist.

reports. Sightings would eventually number in the thousands. Biologist Roy P. Mackal, known for his keen interest in lake monsters, noted in the mid-1970s that "over the years there have been at least 10,000 known *reported* sightings at Loch Ness but less than a third of these *recorded*."

The "Classic" Monster

In July 1930, three local men had a strange experience while fishing from a boat in Loch Ness; one of the witnesses recalled the event in the *Northern Chronicle* (August 27). Noticing "a commotion about 600 yards up the loch," the man saw "a spray being thrown up into the air at a considerable height." So strong was the rush of water, he reported, that it "caused our boat to rock violently." The source of the ruckus seemed to be something large, "traveling at fifteen knots [nautical miles]." Nearly "twenty feet" of it was visible, "standing three feet or so out of the water." The witness said it was "without doubt a living creature" and not "anything normal."

Though this newspaper account brought letters from other readers relating their own or other people's experiences with mysterious animals in the loch, the matter attracted only local attention.

That would change with an incident that took place on the afternoon of April 14, 1933, near Abriachan, a village on the northwest side of Loch Ness. A couple in a passing car spotted a mass of churning water and stopped. Over the next few minutes, they watched an "enormous animal rolling and plunging" out on the loch. The May 2 issue of the *Inverness Courier* ran the story, written by Alex Campbell, who would later claim his own sightings. *Courier* editor Evan Barron called the animal a "monster," and the report attracted some attention around Scotland. Then, as other sightings came in (it seems that the expansion of an old road along the northern shore of the loch had cleared away many natural viewing obstructions), the world eventually took notice. By October, with over 20 reports recorded since the April 14 sighting, the "Loch Ness monster" was born.

Over the years, a clear picture of the monster has emerged. The classic "Nessie" has a long, vertical neck with a small head. Near the end of the neck, some witnesses have reported seeing what looks like a mane of hair, and the head may be horselike in appearance. The long, tapering body may have one, two, or three humps and a long, thick tail. Those who have claimed to see Nessie on land usually report fins, which allow for clumsy forward movement on land but rapid movement in the water. Its color ranges from dark gray to dark brown to black. It surfaces and descends quickly and vertically. It almost always appears when the lake is calm.

A Number of Nessies

Several Nessie experts argue that if the loch is the home of an unusual animal, more than one specimen must exist. Early writers like Gould thought that the lake was home to a single creature, but many feel that the monster must be part of a breeding group. In fact, sightings of more than one creature, though rare, have been made from time to time. The *Scottish Daily Press* of July 14, 1937, for example, told of eight people who observed "three Monsters about 300 yards out in the loch. In the center were two black shiny humps, 5 ft. long and protruding 2 ft. out of the water and on either side was a smaller Monster."

Also, the sizes of the creatures described suggest that there is more than one Nessie. Lengths range from as little as three feet (in rare reports of "baby monsters") to as much as 65 feet! Still, most sightings report creatures between 15 and 30 feet long.

Blurry image, said to be the Loch Ness monster, taken by a tourist at Fort Augustus, Scotland, in 1934.

Land Sightings

Among the thousands of sightings of Nessie, there are a group of reports that teeter on the edge of incredibility—land sightings. Henry H. Bauer, a scientist interested in the Ness mystery, commented:

A considerable but unavoidable embarrassment to the most hardheaded hunters is the existence of a small number of reports of Nessies having been seen on land. In Nessiedom these events have a place that is not unlike that of the 'close encounters of the third kind' in ufology. One is brought squarely up against ... apparently responsible and plausible individuals who insist on ... experiences of the most extremely improbable sort.

Bauer added that such land sightings have been a part of the mystery from the beginning and that they have made belief in Nessies far more difficult.

The most famous of the land sightings took place on the afternoon of July 22, 1933. As they drove down the east side of the loch between Dores and Foyers, Mr. and Mrs. F. T. G. Spicer said they saw a strange animal 200 yards ahead of them. "It did not move in the usual reptilian fashion," Mr. Spicer said, "but with these arches. The body shot across the road in jerks, but because of the slope we could not see its lower parts and saw no limbs." Twenty-five to 30 feet long, it had an elephant-gray color, a bulky body, and a long neck. "We saw no tail," he recalled, "nor did I notice any mouth on what I took to be the head of the creature." The Spicers called the animal, which disappeared into the coarse ferns along the loch, a "loathsome sight."

At around 1 A.M. on January 5, 1934, veterinary student W. Arthur Grant was riding his motorcycle just north of Abriachan when he nearly collided with a strange creature. It lurched across the road, crashed through some nearby plants, and splashed into the loch, disappearing at once. Grant later related: "It had a head rather like a snake or an eel, flat at the top, with a large oval eye, longish neck and somewhat larger tail. The body was much thicker towards the tail than was the front portion. In color it was black or brown and had a skin rather like that of a whale."

The last known land sighting, said to have taken place on February 28, 1960, was of a similarly classic specimen. In mid-afternoon, Torquil MacLeod observed a long-necked animal with flippers through his binoculars. The upper half of the creature was on the shore, with the lower half tapering off into the loch. He watched for nine minutes before it turned and "flopped into the water and apparently went straight down."

Other Oddities

There are plenty of land sightings that do not describe the classic Nessie. Some of them are quite bizarre and give scientifically trained Ness investigators headaches trying to understand them.

In June 1990 *Scots Magazine,* for instance, published Colonel L. Fordyce's account of a land sighting that he and his wife had experienced in April 1932. Driving through the woods one morning along the south side of the loch, they saw "an enormous animal" cross the road 150 yards ahead on its way to the water. "It had the gait of an elephant," he recalled, "but looked like a cross between a very large horse and a camel, with a hump on its back and a small head on a long neck.... From the rear it looked grey and shaggy." It had long, thin legs and a thin, hairy tail. Because it was a year before the Loch Ness monster was publicly recognized, the couple had no clue as to what they were seeing. They thought it was a freak animal escaped from a zoo.

Strange as it may seem, other unlikely beasts have been reported. According to a letter written in 1933 to *The Scotsman* by Patrick Rose, a lake monster "which was a cross between a horse and a camel" had been reported by one of Rose's ancestors way back in 1771. In 1912, from a distance of no more than a few yards, a group of children at Inchnacardoch Bay saw something resembling a long-necked camel enter the loch. A pale, sandy color, it had four legs.

But that was not the end of reports of outlandish beasts in or near the loch. Driving along its northern shore early one April morning in 1923, Alfred Cruickshank reportedly spotted a creature with "a large humped body standing about six feet high with its belly trailing on the ground." Around "twelve feet long," it had a tail of equal length. Its four legs were as thick as an elephant's, and it had large webbed feet. The head was "big and pug-nosed and was set right on the body," and Cruickshank thought the animal looked like an "enormous hippo." Before disappearing into the water, the beast gave a "sharp bark." And in December 1933, a Mrs. Reid claimed to have seen what looked to her like a dark, hairy hippopotamus as it rested on the loch shore.

Land sightings all but stopped after the mid-1930s. But beasts other than the classic monster were still seen in the water. A correspondent writing in the June 7, 1933, issue of the Scottish newspaper *Argus,* for example, claimed that when flying over the loch the week before, he and his companions had seen "in the depths a shape resembling a large alligator, the size of which would be about 25 feet long by four feet wide." Just one year earlier, a woman had spotted a six- to eight-foot animal swimming up the river Ness that she described as a "crocodile." What's more, according to some early nineteenth-century reports, a critter akin to a "great salamander" also appeared in the loch from time to time.

Photographs

Despite these troubling descriptions, the majority of Nessie reports are of a long-necked creature that resembles nothing so much as a plesiosaur, a plant-eating water reptile thought to have become extinct some 65 million years ago. In most Nessie sightings, descriptions vary only in the reported size of the beast.

Hugh Gray, who lived at Foyers on the loch's southeast shore, took the first picture of Nessie on November 13, 1933 (see page 420). He was 200 yards away and about 40 feet above the water. The photograph shows an unclear large object that appears to be moving vigorously. Something extends from its left side, possibly a neck or fin; Gray thought it was the neck, with the head under water. While generally considered genuine, the picture is too fuzzy to settle anything. As J. R. Norman of the British Museum of Natural History said at the time, "I am afraid that the photo does not bring the mystery any nearer to a solution."

The following April, Robert Kenneth Wilson claimed to take what would become the most famous of all Nessie still pictures, the "surgeon's photograph," though Wilson was, in fact, a gynecologist. The widely reproduced picture (actually the first, and better, of two) shows the head and long, curving neck of what appears to be a plesiosaur. Predictably, Wilson's photograph stirred a flurry of argu-

The first photo of Nessie, taken by Hugh Grant in 1933, was deemed authentic, but, like many later photos, it was too blurred to prove anything.

ments for and against its authenticity. Even the date on which it was taken was questioned. And the lesser-known second photograph, showing just the head and a small part of the neck, led some scientists to argue that the animal was small–perhaps a diving bird or an otter.

These arguments appeared minor (and foolish!) when Ness researchers Alastair Boyd and David Martin came forward 60 years later with a bombshell: the "surgeon's photograph" was a fake! A 90-year-old man had confessed this to the two just before his death in November 1993.

It seemed that back in 1933, the *Daily Mail* had hired colorful film-maker Marmaduke "Duke" Wetherell to find Nessie. He hatched a scheme and carried it out with the help of his son, Ian, and Spurling, his stepson. Spurling made a Nessie model–one foot tall and 18 inches long--from molded plastic, wood, and a toy submarine from Woolworth's weighted down with strips of lead. Duke and Ian photographed it in the shallows of a quiet bay in the loch and quickly sank it when they heard someone approach. Wilson was the respectable "middleman" they chose to get the pictures developed. Going along with the hoax to be a good sport, he never imagined that it would be so successful and said little about the photos throughout his lifetime.

The famous 1934 R. K. Wilson photograph of Nessie that, in 1993, proved to be a hoax.

There would be more photographs of the monster. On August 24, 1934, F. C. Adams took an important shot of what looked like a fin attached to an unseen large animal thrashing around in the water near the surface of the loch. Other photographs showed unusual tracks in the water, different from those left by passing boats or other normal lake traffic. Dramatic close-ups of the monster, from pictures taken by Frank Searle (in the early to mid-1970s) and Anthony "Doc" Shiels (1977), have appeared in many magazines and books, but serious researchers consider them hoaxes.

Underwater photographs taken in 1972 and 1975 looked like more evidence in Nessie's favor. On the night of August 7, 1972, investigators of the Massachusetts-based Academy of Applied Sciences and the Loch Ness Investigation Bureau were patrolling the waters near Urquhart Bay. One boat contained a sonar (sound wave) device; another held strobe and camera equipment. (A strobe is an instrument that employs a flashtube for high-speed lighting that is sometimes used in photography.) At 1 A.M. sonar picked up an unidentified target about 120 feet away, within range of an underwater strobe camera but apparently above or below its beam, as nothing was captured on film. Forty minutes later, two large objects, 20 to 30 feet long and about 12 feet apart, were tracked, as were some salmon fleeing before them. The traces stopped after a few minutes.

Academy investigators took the film from the underwater camera to the Unites States to have it developed at the head office of photographic-equipment company Eastman Kodak. Two frames showed what looked like a big flipper attached to a body of rough texture. A third showed blurry images of two objects (just where the sonar echoes had placed them). The clearer of these objects suggested a classic Nessie: long neck, bulky body, and fins.

To improve the flipper photographs, which were murky because of the loch's peat-sogged water, the researchers took them to the Jet Propulsion Laboratory (JPL), where state-of-the-art photo studies were regularly performed for official, military, and scientific agencies. There, through a standard computer-enhancement technique, much of the graininess of the original photographs was removed. These clearer pictures would appear in *Nature* magazine and elsewhere. "This technique has proven to be a real tool," academy investigators wrote. "It has been used to clarify images from space probes, in forensics to help identify fingerprints, and in medical research to classify human chromosomes." And "it cannot create patterns where there are none."

The flipper in the pictures appeared to be four to six feet long. Those who studied the sonar records agreed that the objects detected by the sound waves were also the subjects of the photographs. To British television newsman Nicholas Witchell (author of a popular book on the monster), this "coincidence of the sonar and the photography ... presented indisputable proof of the animal's presence. The one cross-checked ... the other. Here was the breakthrough." Scientists and journalists did take notice. Even *Time* magazine, which has historically avoided covering strange claims of any kind, stated, "Now the skeptics may have to re-examine their doubts."

In June 1975 the academy team produced even more amazing evidence. This time it was two dramatic pictures, taken about seven hours apart on the morning of June 20. The first showed, according to *Technology Review,* the "upper torso, neck and head of a living creature." Again Ness's murky water made this identification less than certain; still, little imagination was needed to detect the features. Even more startling was the second photograph. It appeared to be the monster's head, just five feet from the camera! It was horselike and even had the small horns that some observers had reported. According to the investigators, "Measurements indicate the 'neck' to be about one-and-one-half feet thick, the 'mouth' nine inches long and five inches wide, and the horn on the central ridge six inches

Even more startling was the second photograph. It appeared to be the monster's head, just five feet from the camera!

long." The photograph quickly became known as the "gargoyle head" picture.

At first, scientific response to this newest piece of Nessie evidence made believers hopeful. Because of the pictures, zoologists from Washington D.C.'s Smithsonian Institution, the Royal Ontario Museum, Harvard University, the New England Aquarium, and other important institutions either outright supported Nessie's existence or declared that it was now a real possibility. But scientists from London's Natural History Museum expressed doubt, which would spread. Because a portion of the neck near the head in the first 1975 photograph was not visible (lost in the shadows, according to believers), they did not feel that the photo indicated the creature existed. "This probably should be interpreted as two objects," they wrote; "conceivably various floating objects could assume this form."

Certainty about the academy photographs began to erode. In 1984 the popular-science magazine *Discover* objected to the computer enhancements performed on the 1972 flipper photographs, claiming that secret retouching had turned "grainy and indistinct" images into false evidence of an unknown animal. But the academy had not tried to hide anything, clearly admitting to computer assistance with the photographs. And Allan Gillespie, who worked on the pictures at JPL, knew that the academy was not trying to make something out of nothing. "The outline of the flipper is visible in the original," he stated.

Then Adrian Shine of the Loch Ness and Morar Project insisted that the 1975 "gargoyle head" was really a rotting tree stump. Shine brought the stump to the loch's surface and photographed it and displayed the resulting print next to the head picture, remarking on the similarities. Most observers could find no resemblance. And Tim Dinsdale, one of the most famous of Nessie's hunters, suggested a different explanation for the object in the photo: turning its angle would reveal a car or truck's engine block with exhaust pipe. Such things were often tossed into Urquhart Bay to anchor boats.

So what first seemed clear evidence of Nessie's existence only brought more questions in the end. Still, in its favor, the academy photographs look to most observers more like animal parts than anything else.

> **IS LOCH NESS A SUITABLE HOME FOR NESSIE?**
>
> More evidence challenging the existence of Nessie surfaced in late 1993. A team of scientists studying the ecology of Loch Ness revealed that the lake has such a small fish population that it could not possibly support a family of monster-sized predators.

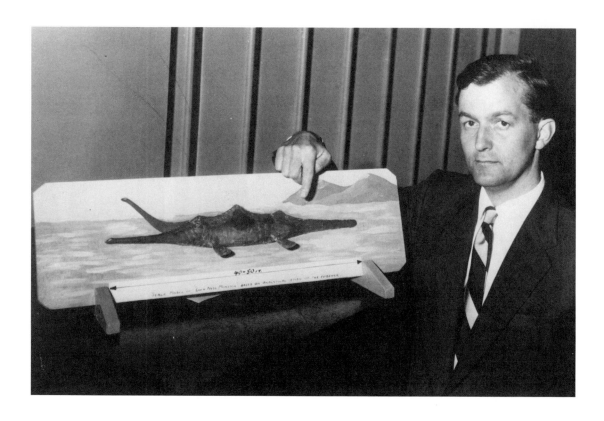

Former Royal Air Force pilot Tim Dinsdale displays a model he made of the Loch Ness Monster.

Film Clips

Malcolm Irvine of Scottish Film Productions took the first motion-picture film of Nessie on December 12, 1933, in Urquhart Bay. Two minutes long, it showed a long, dark object that barely protruded above the water's surface 100 yards away. Like most Nessie films that would come later, it was intriguing but would provide no answers.

The whereabouts of one film that might have held some answers is unknown–if it ever existed at all. Said to have been taken sometime in the 1930s by a London physician named McRae, it supposedly showed several minutes' footage of a three-humped creature with a cone-shaped head, two horns, a stiff mane down a long neck, and—at one point, when the animal rolled over—a flipper. According to McRae's close friend, artist Alastair Dallas, the doctor feared ridicule and decided not to release the film. Nessie investigator F. W. Holiday learned about it from Dallas in the early 1960s.

The coming decades would bring many more Nessie films. In 1977 biologist Mackal reviewed 22 of them. He dismissed nearly half because of poor quality. Six involved mistaken identifications of known objects. And five contained "positive evidence." Among the most important of these was one taken by Dinsdale.

On April 23, 1960, on the last day of a six-day watch of the loch, Dinsdale was sitting in his car on a hill near Foyers on the eastern shore. Suddenly he spotted an unusual-looking, motionless object two-thirds of a mile across the water. With his binoculars he saw a "long oval shape" that "had fullness and girth and stood well above the water." "It began to move," he related. "I saw ripples break away from the further end, and I knew at once I was looking at the extraordinary humped back of some huge living creature!" He filmed it with a 16-mm camera for four minutes as it swam away, partially underwater. Running out of film, he stopped the camera in the hope that the creature would show its head and neck. It did not.

Wisely, Dinsdale later filmed an associate sailing a boat in the same direction that the creature had taken. In 1966, when Britain's Joint Air Reconnaissance Intelligence Center (JARIC) studied the first film, it used the second reel for comparison. Center members were able to closely guess the boat's size and speed. Then, turning to the first film, they determined that the object was definitely not a boat (a favored explanation), but "probably ... an animate object." They also figured that the hump was between 12 and 16 feet long and about three feet above the water. The object was moving around 10 mph. The Dinsdale film is considered a major piece of evidence of Nessie's existence, and in the years since no one has seriously challenged the JARIC study.

Sonar

Sonar has tracked Nessie-like targets many times. The first tracings were recorded in 1954, when a commercial vessel noted a large moving object passing 480 feet below it. And between 1968 and 1970, using sonar devices from shore and on boats, D. G. Tucker of the University of Birmingham and his associates tracked 20-foot living things that swam and dived near the bottom and sides of the loch. Sometimes they were tracked in groups; one time the group counted from five to eight members. Their behavior, speed, and size convinced Tucker that they were not fish.

Investigators have continued to make sonar trackings of creatures in the loch. The Loch Ness and Morar Project and technology and electronics companies have been behind many of them. In 1987 the most involved and well known of these, Operation Deepscan, brought more than 20 vessels to the loch surface for a three-day sonar sweep, between October 8 and 10. Though they covered only the lake's southern half, ten contacts were recorded.

Explanations

The idea that the loch's monsters are large eels is one unusual but believable explanation. Elephant seals or sea cows (plant-eating water mammals like the dugong, or manatee) have also been named as unusual but believable candidates. The only problem is that none of these look much like what witnesses report or what the photographs and films seem to show.

Of the extraordinary explanations the most popular is that the animals are surviving plesiosaurs that have adapted to Ness's cold temperatures. But a few investigators, like Mackal, favor the zeuglodon theory, named for a primitive, snakelike whale also thought to have become extinct long ago. No single theory satisfies all the data, and investigators of the Loch Ness mystery disagree even on what "the data" is. Meanwhile, the search continues.

Sources:

Bauer, Henry H., *The Enigma of Loch Ness: Making Sense of a Mystery,* Urbana, Illinois: University of Illinois Press, 1986.

Binns, Ronald, *The Loch Ness Mystery Solved,* Buffalo, New York: Prometheus Books, 1984.

Dinsdale, Tim, *Loch Ness Monster,* fourth edition, Boston: Routledge and Kegan Paul, 1982.

Ellis, William S., "Loch Ness: The Lake and the Legend," *National Geographic* 151,6, June 1977, pp. 759-779.

Holiday, F. W., *The Dragon and the Disc: An Investigation into the Totally Fantastic,* New York: W. W. Norton and Company, 1973.

Holiday, F. W., *The Great Orm of Loch Ness: A Practical Inquiry into the Nature and Habits of Water-Monsters,* New York: W. W. Norton and Company, 1969.

Mackal, Roy P., *The Monsters of Loch Ness,* Chicago: The Swallow Press, 1976.

Witchell, Nicholas, *The Loch Ness Story,* Baltimore: Penguin Books, 1975.

CHAMP

C hamp is, at least according to some, Lake Champlain's version of the Loch Ness monster: a large, long-necked animal that looks like a plesiosaur, a water reptile that became extinct some 65 million years ago (also see entry: **Lake Monsters**). But in reality, eyewitness accounts describe many different creatures. There have been more than 300 recorded sightings of Champ, according to Joseph Zarzynski, who has investigated and written about the monster.

The popular belief is that Samuel de Champlain, the French explorer after whom the lake was named, was the first white man to see the monster. He was supposed to have mentioned it in a 1609 account of his travels on the Saint Lawrence and other rivers. But he only refers to large fish in his report. His description suggests garfish, which can still be found in Champlain today.

Early Sightings

The story of Champ really begins, more or less, in 1873 when the first known newspaper story about a monster in Lake Champlain appeared in the *Whitehall Times* on July 9. In this and other early accounts, the monster was described as a different animal from our modern version—a giant serpent and not a plesiosaur.

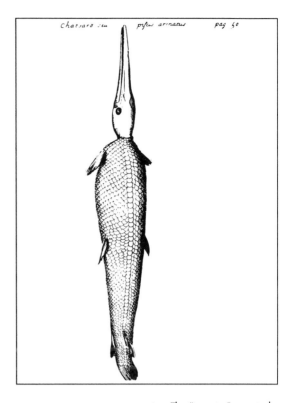

The "monster" reported by Samuel de Champlain; from *Historia Canadensis* (Paris, 1664) by Father François De Colux.

According to the *Times,* a railroad work crew laying tracks on the lakeshore near Dresden, New York, "saw a head of an enormous serpent sticking out of the water and approaching them from the opposite shore." So terrified were the men that they stood paralyzed for a while before scattering. The monster turned toward the open water and then:

As he rapidly swam away, portions of his body, which seemed to be covered with bright silver-like scales, glistened in the sun like burnished [polished] metal.... From his nostrils he would occasionally spurt [streams] of water above his head.... The appearance of his head was round and flat, with a hood spreading out from the lower part.... His eyes were small and piercing, his mouth broad and provided with two rows of teeth, which he displayed to his beholders.

As he moved off ... portions of his body appeared above the surface of the water, while his tail, which resembled that of a fish, was thrown out of the water quite often.... A quarter-mile into the lake, the creature sank suddenly out of sight.

Within a few days of the Dresden event, farmers were complaining of missing livestock. Tracks and other marks on the ground suggested that something had dragged the animals into the lake. In caves along the waterside, local residents claimed that "bright and hideous looking eyes" could sometimes be seen in the darkness. A few days later a young farmer saw the serpent in a lakeside marsh, with something that looked like a turtle in its mouth. He fired on it, and the creature disappeared into the water.

Other sightings and livestock kills followed, and search parties prowled the shoreline and surrounding farms. In early August the small steamship *W. B. Eddy* struck the serpent and nearly overturned. On August 9 the crew of the *Molyneaux* believed it had the monster trapped in the thick weeds of Axehelve Bay. Though no one could actually see it, guns were fired into the thickets. The *Whitehall Times* of August 13 printed one witness's account of the results:

> Our ears were greeted by a most unearthly noise.... The order was given (by Captain Belden) to steam away as the head of the mammoth snake appeared through the tangled vines and brushwood. The greyish hood upon his head flopped backwards and forwards.... Great ridges of silver appeared above the surface of the water, undulating and scintillating in the bright sun like the highly polished surface of a warrior's silver helmet.... His eyes resembled two burning coals, fairly snapping fire, as its rage increased, while the rows of long and formidable teeth, pearly white and wicked looking, sent an indescribable thrill through us, which we shall never forget. The body seemed to be about 18 to 20 inches thick in the middle, and 36 to 40 feet long.

As the *Molyneaux* retreated, more shots were fired at "the great moving, waving mass of silver." As the creature "lashed the water with his fishlike tail and gave great spasmodic, powerful lurches with his broad flat head," the crew knew that their gunfire was taking its toll. Finally, two well-placed shots from just 25 feet away hit their mark, and "streams of red blood spurted from its head." After one last "spasmodic twist," the monster "disappeared beneath the red sea of blood."

Attempts to raise the body proved unsuccessful. Searchers hoped to collect the $50,000 reward offered by entertainer P. T. Barnum, who

LAKE CHAMPLAIN

Formed about 10,000 years ago out of melting glaciers, Champlain is a deep, cold-water, 109-mile lake. It runs along the New York/Vermont border and into Quebec, where it drains into the Saint Lawrence River and finally the North Atlantic Ocean. The largest North American body of water outside the Great Lakes, it is the home to some 80 different species of fish.

wanted the "hide of the great Champlain serpent to add to my mammoth World's Fair Show."

All these events reportedly took place in the Dresden area at the southwestern edge of the lake. When the monster (or a relative) returned a few years later, it had moved to Champlain's northwestern side, near Plattsburgh.

On July 31, 1883, the *Plattsburgh Morning Telegram* reported that Clinton County Sheriff Nathan H. Mooney had seen an "enormous snake or water serpent ... 25 to 35 feet in length." Three years later, beginning in the summer of 1886, sightings were recorded almost daily from practically every part of the lake. One man fishing near Plattsburgh claimed to have hooked what he first thought was a huge fish. But when he and three other witnesses finally saw its head they realized it was a "horrible creature." The line snapped, and the terrifying catch fortunately disappeared underwater!

Sightings continued into the next year. At two o'clock one morning in May 1887, for instance, a farm boy heard strange noises and went to the lakeshore. A mile out in the water he spotted a big serpent "mak-

List of Champ sightings, beginning with Samuel de Champlain.

ing noises like a steamboat." In several other cases witnesses claimed that the creature had threatened them, swimming fiercely at them and forcing their quick retreat. In one gripping account reported in early July, picnickers from Charlotte, Vermont, noticed the beast from their spot near the lake. Seventy-five feet long and as big around as a barrel, it bore down on the group until several women screamed. It then turned around and swam off.

Shortly afterward, this incredible report was published in the *Plattsburgh Morning Telegram:*

> The sea serpent ... has left the lake and is making his way overland in the direction of Lake George [New York]. He was seen last night about five o'clock by a farmer driving to his barn with a load of hay. Chancing to look behind him ... he saw ... gliding along like a snake with its head raised about four feet from the ground ... an immense monster anywhere from 25 to 75 feet in length, ... covered with scales.

A few sightings of the monster, though, revealed unserpentlike details. In September 1889 a group of fishermen chased the creature; it had, they reported, "many large fins" (*Essex County Republic,* September 26). In the summer of 1899, a witness reported seeing a 35-foot-long serpent with a broad, flat tail raised a few feet above the water (*Plattsburgh Republican,* August 5). This feature—if truly observed—makes the animal a mammal, not a reptile. Perhaps, as some have suggested, it was a zeuglodon, a snakelike whale thought to have become extinct some 20 million years ago.

The Twentieth-Century Monster

If anything, the confusion about Champ's identity has grown in the twentieth century. Indeed, the plesiosaurlike Champ would become the "classic" monster only in the 1970s. Elderly residents of the Champlain area would claim to recall seeing plesiosaurlike creatures—with long necks and bulky reptile bodies—earlier in the century. But investigators would suspect that their memories had been influenced by the monster's more recent image.

In fact, twentieth-century accounts before 1970 seemed to describe a variety of creatures. Huge snakes often were reported, as in the century before. Some mentioned scales, which snakes do not have. Others seemed to describe large fish, perhaps sturgeons. And in a few cases, sightings suggested (nonprehistoric) whales and eels.

In the summer of 1899, a witness reported seeing a 35-foot-long serpent with a broad, flat tail raised a few feet above the water.

In one of the very rare twentieth-century land sightings–this one made from a passing car in the spring of 1961—Thomas E. Morse reported what "appeared to be a monstrous eel with white teeth that raked rearward in the mouth." It was resting on the shore of Champlain's North West Bay.

A Champ sighting that took place in the summer of 1970 is particularly interesting. It involved separate witnesses and shows just how differently people can view the same thing. The *Plattsburgh Valley News* of August 9, 1978, published two accounts written by the witnesses, who had not spoken in the eight years since their shared sighting. Richard Spear recalled seeing the creature with his 13-year-old daughter as the two of them sat atop a ferry heading toward the Essex, New York, shore. Another ferry witness was Happy Marsh.

In her account, Marsh (who also claimed to have seen an identical animal in 1965 or 1966) described "a large snakelike creature, swimming with her head above water, held as snakes do, with coils behind." She guessed that it was "between 18 and 20 feet long.... It was black, and swimming slowly. Her head was three feet long, wrinkled like a raisin, with a small ridge down the back."

Spear, on the other hand, described an animal that looked more like a plesiosaur. It was barrel-shaped and had "two 'bumps' ... each rising to about three feet above the surface and four feet in length." When his daughter used binoculars, she saw its head, which she said looked like a horse's. The creature was "dark brownish-olive" in color.

Champ vs. the Loch Ness Monster

Reviewing Joseph W. Zarzynski's 1984 book *Champ: Beyond the Legend,* Henry H. Bauer, a leading expert on the **Loch Ness monster,** complained that the author's effort to link the two animals made little sense. He stated: "The listed sightings [of Champ] include a goodly number of descriptions as 'snakelike,' which has never been said of Nessie; smooth skin is reported whereas Nessie's is rough, warty; eyes are featured several times, and fins and manes, which are almost totally lacking in reports from Loch Ness."

Still, some clear reports of a Ness-like animal in Lake Champlain do exist. The "prehistoric monster" that Orville Wells saw in Champlain's Treadwell Bay in 1976 (and later sketched) resembled the creatures of Loch Ness. Several other witnesses have specifically said that what they saw looked like a "dinosaur." And the Mansi photograph–the most

Joseph Zarzynski, director of Lake Champlain Phenomena Investigation, readying a sonar tripod for use in his search for Champ.

important piece of evidence that Champ is real—clearly shows a plesiosaurlike animal.

The Mansi Photograph

In early July 1977, a Connecticut couple, Anthony and Sandra Mansi, were vacationing in Vermont. Just past Saint Alban's Bay and somewhere near the Canadian border, they stopped so that their two small children could play in Champlain's water. They parked their car and walked 100 to 200 feet across a field, then went down a six-foot bank to the waterline. As the children waded near the shore, Mansi went back to the car for his sunglasses and a camera.

Some moments later Mrs. Mansi noticed bubbles in the water about 150 feet away. Soon a huge animal with a small head, long neck, and humped back rose to the surface. Mrs. Mansi thought it looked prehistoric.

By this time her husband had returned, and he, too, watched the thing with growing alarm. He and Sandra called the children (who were unaware of what was happening in the water behind them and never saw the creature). Anthony tossed the camera to Sandra, telling her to take a picture. She took one photograph before the animal sank—it did not dive—under the water. The sighting lasted between two and four minutes.

Afraid that they would be ridiculed, the Mansis kept their experience a secret. They placed the photograph, which turned out quite well, in a family album! In time they lost the negative. Eventually Mrs. Mansi showed the picture to friends, and by 1980 rumors of it had reached Zarzynski, a social studies teacher in Wilton, New York. He won the couple's trust and began an investigation.

He showed the photograph to a number of experts, including George Zug at the Smithsonian Institution's Museum of Natural History. Zug said it resembled no known animal in the lake or elsewhere. Roy Mackal, biologist and vice president of the International Society of Cryptozoology, also examined it. Soon afterward B. Roy Freiden of the University of Arizona's Optical Sciences Center made a careful study of the photograph.

Freiden determined that the photograph had not been tampered with—no one had imposed Champ's image over a picture of the lake. The wave patterns around the animal also suggested that what Mrs. Mansi had observed was true: the "object" had come up from under the surface instead of moving along the water (as would be the case if it were, say, an artificial device being pulled by a rope).

Freiden could not make out the object's exact size because the photograph provided no other features for comparison. It did not, for example, show the shoreline, which would have helped investigators determine the size of the object and its distance from observers. But University of British Columbia oceanographer Paul LeBlond found another method for at least guessing at the size: by measuring the length of the waves around the object. Using wind speed and the distance of open water over which it blew, LeBlond figured that the waves in the photograph were between 16 and 39 feet long. Then that part of the "unknown object" above water was compared with the waves. It was determined that its length ranged from 24 to 78 feet!

If the enormous "object" in the photograph was artificial, what an extraordinary—and expensive—hoax! And if it was a trick, why did Mrs. Mansi wait for more than three years before allowing it to become public? And what is more, why would she take only *one* photograph? In all the time since the sighting, no evidence has emerged suggesting a hoax. The Mansi photograph remains a mystery. And it also supports the often-questioned link between Champ and Nessie.

The Continuing Investigation

In the 1970s Zarzynski formed the Lake Champlain Phenomena Investigation (LCPI). The LCPI interviews witnesses and searches for historical references to Champ. It also keeps a close eye on the lake—visually and electronically (so far with little results).

Nonetheless, the creature's existence may be based on hoaxes or mistaken identifications. Certainly the many different descriptions of the beast suggest that Champ is wholly imaginary or a catchall name for a variety of animals—some known but out of place, some unknown. Because Champlain is linked to the ocean, it is possible that oversized sea animals wandering in and out are behind the Champ legend. Better evidence is needed to get to the bottom of the mystery at Lake Champlain.

Sources:

Coleman, Loren, *Mysterious America,* Boston: Faber and Faber, 1983.
Zarzynski, Joseph W., *Champ: Beyond the Legend,* Port Henry, New York: Bannister Publications, 1984.

MORAG

On April 3, 1971, Ewen Gillies, a lifelong resident of a house overlooking Loch Morar, saw the creature for the first time. His 12-year-old son John had noticed it a few minutes before, while walking down a road near the shore. He told his father, who stepped outside on the clear, sunny morning and looked out at the lake. Not quite half a mile away, a huge animal lay in the water, its three- or four-foot neck pointed straight up and curving slightly at the top. It was hard to tell where the neck ended and the head began. Two or three humps moving up and down slightly ran along its back. The skin was black and shiny. The creature was around 30 feet long.

Gillies went into the house to get a camera. He took two pictures from an upstairs window just before the creature lowered its head, straightened its body, and sank below the surface. The pictures did not turn out, but no one accused Gillies of concocting the story, for Loch Morar's monster—Morag—had been seen before.

Loch Morar, Inverness, Scotland, haunt of Morag.

Early History

Morag comes from the Gaelic word *Mhorag,* the name of a monster that was once believed to be the spirit of the loch. A shape-changing mermaid, it was thought to mean death if glimpsed by a member of the Gillies clan, who had lived in the area for centuries. As time passed and people migrated from this wild, remote region of Scotland, the old folklore faded away. Morag was then considered a strange—but not supernatural—creature seen by some but rarely discussed.

Because of Morag's rich folkloric past, researchers have had difficulty tracing individual sightings back before the late nineteenth century. Interviews with elderly residents, however—like those by investigator Elizabeth Montgomery Campbell in the 1970s—have revealed that locals remember seeing the beast when they were young.

LOCH MORAR

Loch Morar is 70 miles southwest of Scotland's much more famous Loch Ness. Eleven miles long and a mile and a half across at its widest point, it is separated from the sea by only a quarter of a mile and is one of Scotland's deepest lakes. According to reports that go back decades (and perhaps even centuries), it is the home of monsters very much like those reported at Loch Ness and other Scottish and Irish lakes.

Witnesses

Lady Brinckman was one such witness; she lived on an estate near the loch around the turn of the century. She recalled an incident in the summer of 1895 in an unpublished memoir written decades later. She related that "one evening, it was getting towards dinner time and I was sitting looking back, when suddenly, I saw a great shape rise up out of the loch, a good way off." Asking her two male companions if it were a boat, one replied, "It'll just be the monster." He said such sightings of the monster were not uncommon.

Folklorist R. Macdonald Robertson recorded a story from Alexander Macdonnell describing an event that took place early in the century. "Some years ago, we were proceeding one morning down the loch in the estate motor launch from Meoble to Morar pier with some schoolchildren and other persons on board," recalled the witness. "As we were passing Bracarina Point, on the north side, some of the children [excitedly] shouted out: 'Oh look! What is that big thing on the bank over there?' The beast would be about the size of a full-grown Indian elephant, and it plunged off the rocks into the water with a terrific splash." Robertson noted that a number of other reliable witnesses had seen Loch Morar's monster. One observer described the typical sighting, "a huge, shapeless, dark mass rising out of the water like a small island."

In September 1931 young Sir John Hope (Lord Glendevon) had an odd experience in the loch. While it involved no direct sighting, it clearly suggested the presence of some huge unknown animal. He, his brother, a friend, and a local guide had gone out on a boat to fish in a deep part of Morar. Hope, who was holding a long trout rod, felt something grab his line, dragging it "directly downwards at such a pace that it would have been madness to try and stop it with my fingers. In a very few seconds the whole line, including the backing, had gone and the end of the rod broke." Hope said that whatever took the bait was "something ... heavier than I have experienced before or since."

What could the creature have been? A salmon—if one that size even existed—would have traveled parallel to the surface of the water instead of making a steep vertical descent. A seal might swim downward, but no seals were known to live anywhere near Loch Morar. But such descents are described in a great many lake monster sightings. Glendevon recalled that when he and his companions asked their guide what the animal could have been, the man "mumbled something and said he thought we had better go home." Glendevon suspected that the guide knew more than he was telling.

The Loch Morar Survey

After 1933, the year the first photograph of the Loch Ness monster appeared, Morag also received some attention. A few witnesses came forward and described sightings of large, fast-moving humps in the

"The beast would be about the size of a full-grown Indian elephant, and it plunged off the rocks into the water with a terrific splash."

Whatever took the bait was something heavier than Sir John Hope had ever experienced.

water or of long-necked creatures, usually about 30 feet long, in Loch Morar. In February 1970 several members of the Loch Ness Investigation Bureau formed the Loch Morar Survey. Over the next few years they launched investigations as their limited time and funds permitted. On July 14, 1970, one of them, marine biologist Neil Bass, spotted a "hump-shaped black object" in the lake. He called to his associates, but the hump disappeared before they had a chance to see it.

Bass reported that "within half a minute," though, "a disturbance was witnessed by all of us ... followed by radiating water rings which traveled to form a circle, at maximum 50 yards in diameter." It was made by something very large. Bass believed that the unknown "object" was a large living creature unfamiliar to him.

Perhaps the most dramatic Morag event recorded took place on August 16, 1969. It was also the only sighting ever to be reported in newspapers across the world. It occurred as two local men, Duncan McDonell and William Simpson, were on their way back from a fishing trip at the north end of the loch. It was just after 9 P.M. and the sun had gone down, but there was still plenty of light.

McDonell, who was at the wheel, turned around after hearing a splash behind them. To his amazement, a creature was coming directly toward them, at about 20 to 30 mph! Within seconds it struck the side of the boat, then slowed down or stopped. Though McDonell felt the collision was an accident, he was still afraid that the huge animal would overturn them. He grabbed an oar and tried to push it away. Meanwhile Simpson had rushed into the cabin to turn off the motor. He returned with a rifle and fired a single shot at the beast, seemingly with no effect. It moved away and sank out of sight. The incident lasted five minutes.

When interviewed by members of the Loch Ness Investigation Bureau, the two agreed that the creature had been some 25 to 30 feet long, with rough, dirty brown skin. Three humps, about 18 inches high, stood out of the water, and at one point McDonell had spotted the animal's snakelike head just above the surface.

Theories

Loch Morar lies in a deep valley carved out by glaciers on Scotland's west-central coast. Twelve thousand years ago, as the ice retreated, ocean water is believed to have entered the lake, bringing with it a wealth of sea life. Even after the saltwater retreated, for a few thousand years the sea animals in the loch might have found it fairly easy to return to their ocean home. For back then, sea level at high tide would have been within a few feet of loch level.

There is no doubt that Loch Morar has enough food—fish, plankton, and other living matter—to feed a population of large animals. It is one of nine Highland lakes with "monster" histories and reports. (Besides Ness, the others are Oich, Canisp, Assynt, Arkaig, Shiel, Lochy, and Quoich.) Most sightings at Morar and elsewhere describe creatures that resemble the supposedly long-extinct plesiosaur. If such animals survive, however (and there is no evidence to support this in fossil records), they would have had to adapt to far colder water temperatures than their ancestors were used to. Roy P. Mackal, a biologist with a keen interest in lake monsters, argues that Morag, Nessie, and their relatives are zeuglodons: primitive, snakelike whales generally thought to have become extinct some 20 million years ago.

Is the idea of surviving giant prehistoric reptiles and mammals too fantastic? The lake creatures that people report seeing often look like plesiosaurs and zeuglodons. It seems equally fantastic to explain away these sightings as something more acceptable—like sharks, seals, or seaweed, which these "monsters" simply do not resemble. In the meantime, the mystery at Loch Morar continues.

Sources:

Campbell, Elizabeth Montgomery, and David Solomon, *The Search for Morag,* New York: Walker and Company, 1973.

Holiday, F. W., *The Dragon and the Disc: An Investigation into the Totally Fantastic,* New York: W. W. Norton and Company, 1973.

Mackal, Roy P., *The Monsters of Loch Ness,* Chicago: The Swallow Press, 1976.

OGOPOGO

Around eight o'clock on a pleasant morning in mid-July 1974, a teenage girl was swimming just offshore in Lake Okanagan, located in the southern interior of British Columbia, Canada. She was heading for a raft a quarter of a mile from the beach and was only three feet from it when a huge, heavy something bumped against her legs. Surprised and frightened, she grabbed for the raft and climbed aboard.

From there she looked into the clear water and saw a strange animal 15 or 20 feet away. "I could see a hump or coil which was eight feet long and four feet above the water," she told J. Richard Greenwell of the International Society of Cryptozoology more than a decade later. "It was traveling ... away from me ... and it swam very slowly.... Five to 10 feet behind the hump, about five to eight feet below the surface, I could see its tail. The tail was forked and horizontal like a whale's, and it was four to six feet wide. As the hump submerged, the tail came to the surface until its tip poked above the water about a foot." Soon the girl lost sight of the creature. The whole experience had lasted just four or five minutes.

Another witness, Mrs. B. Clark, told Greenwell that the animal was a "very dull dark gray" color and moved in an undulating or wavy manner. She had the "impression that the head joined the body without a neck—like a fish or snake....This thing looked more like a whale than a fish, but I have never seen a whale that skinny and snaky-looking before."

The Zeuglodon

But in fact, such a whale existed, at least at one time. Evidence of the animal has been found in fossils, but these are 20 million years old or more. Known to zoologists and paleontologists, the creature is the *Basilosaurus,* or zeuglodon. For decades, something very much like it has been reported in Lake Okanagan. Since 1926 the animal has been called Ogopogo (the word taken from a song).

Despite its silly name, Ogopogo is one of the most credible of the world's lake monsters. Reports of it are strikingly similar. And they do not include the giant serpents found in folklore or the plesiosaurlike creatures made famous at Loch Ness. What is more, zeuglodons are known to only a few people, mainly paleontologists and cryptozoologists.

American Indians living on or around Lake Okanagan were familiar with Ogopogo long before the white settlers came to the area. *Naitaka,* the serpentlike creature, figured in many of their supernatural legends. Around 1860, when the first white settlers arrived in the Lake Okanagan area, they too began to suspect that strange animals lived in the water. One early sighting by settlers took place in the mid-1870s, when two witnesses on opposite sides of the lake watched a long, snakelike creature swim against the wind and current. At first both observers had mistaken the object for a log. Over the years many other witnesses would describe "logs that came alive."

By the 1920s hunting parties from Canada and the United States scoured Okanagan, hoping to kill a specimen. Sightings continued on and off over the decades. One of the more impressive accounts took place on July 2, 1949, in the early evening, when a party aboard an off-shore boat saw a strange animal 100 feet away. It had a "forked" horizontal tail like a whale and moved its snakelike body in an undulating manner (reptiles move from side to side, not up and down). About 30 feet of a smooth, dark back was visible. The head was under water, perhaps feeding. Another witness saw the creature from land.

More Sightings

A 1967 sighting by nearly 20 people at Okanagan's southern tip made the whale identification even stronger: "It had a head like a bucket and was spouting water," one witness said. On July 30, 1989, when Ogopogo appeared 1,000 feet away from an investigative team of the British Columbia Cryptozoology Club, John Kirk got a clear view through a telescope. "The animal's skin was whale-like," he reported.

Looking at more than 200 Ogopogo reports collected by Mary Moon, University of Chicago biologist Roy P. Mackal put together this general description: "The animals look most like a log, elongated, serpentine, no thickened body centrally, about 12 meters [40 feet long], although a range of smaller sizes has been reported and a few larger, up to say 20 meters [70 feet]." He noted that the skin was generally smooth and dark green, brown, or black in color, although a few plates or scales had been reported by close observers. "Most of the back is smooth," he added, "although a portion is saw-toothed, ragged-edged, or serrated. Sparse hair or hair-bristle structures are reported around the head, and

Ogopogo, monster of British Columbia's Lake Okanagan and star of "Ogopogo Days."

in a few cases a mane or comblike structure has been observed at the back of the neck."

To Mackal, these features "fit one and only one known creature"—the zeuglodon. He noted that identical animals had been reported off the coast of British Columbia and in other Canadian lakes. Mackal suggested that Ogopogos were freshwater-adapted versions of the prehistoric zeuglodon, which lived in the oceans.

Since there are no convincing photographs of Ogopogo and no sonar (sound wave) traces or other evidence from instruments, the case for Ogopogo rests entirely on eyewitness testimony.

Sources:

Conklin, Ellis E., "Ogopogo Brouhaha," *Seattle Post-Intelligencer,* March 7, 1991.

Greenwell, J. Richard, "Interview: The Lady of the Lake Talks about Ogopogo," *The ISC Newsletter* 5,6, summer 1986, pp. 1-3.

Mackal, Roy P., *Searching for Hidden Animals,* Garden City, New York: Doubleday and Company, 1980.

WHITE RIVER MONSTER

From about 1915 through the early 1970s, residents of Newport, in northeastern Arkansas, reported seeing a "monster" in the White River, which flows through the town. Sightings were not continuous but tended to occur in bunches. In July 1937, for example, a number of local people saw either strange disturbances in the water or caught glimpses of the creature.

One witness was Bramblett Bateman, who gave sworn testimony that he had seen, on or around July 1, "something appear on the surface of the water." Because he was 375 feet away from the creature, he could not make out its full length or size but guessed that it was around "12 feet long and four or five feet wide." Nor could he make out its head or tail. The animal remained in its position for about five minutes. In later sightings, Bateman saw the creature "move up and down the river."

Jackson County Deputy Sheriff Z. B. Reid was with Bateman when the creature appeared later that July day. They saw "a lot of foam and bubbles coming up in a circle about 30 feet in diameter some 300 feet" from them. Then, many feet farther up the river, the creature surfaced. To Reid, it looked like a "large sturgeon or cat fish. It went down in about two minutes."

The next widely reported series of sightings occurred in June and July 1971. One witness reported seeing a "creature the size of a boxcar thrashing." He added that "it looked as if the thing was peeling all over, but it was a smooth type of skin or flesh." Another witness took a blurry photograph of a large surfacing form on June 28. The witness also described the monster's roar, a combination of a cow's lowing and a horse's whinny. Other observers also reported the roar. And on those rare occasions when the animal's face was briefly seen, it was said to have a jutting "bone" on its forehead.

A Frightening Experience

Ollie Ritcherson and Joey Dupree's brush with the White River monster was the most frightening. They were cruising near Towhead Island looking for the creature when their boat collided with something. Their vessel rose in the air on the back of some huge animal that they were not clearly able to see. The pair had come to the site because two weeks earlier huge tracks leading to and from the river had been found on the island. Each of the three-toed tracks was 14 inches long

They were cruising near Towhead Island looking for the creature when their boat collided with something. Their vessel rose in the air on the back of some huge animal that they were not clearly able to see.

Could an elephant seal, like the one shown here, have wandered away from its normal habitat and adopted the White River as its home?

and eight inches wide, with a large pad and another toe with a spur extending at an angle. There was evidence, in the form of bent trees and crushed plants, that a large animal had walked on the island and even laid down there.

To biologist Roy P. Mackal, the case of the White River monster seemed "a clear-cut instance of a known aquatic animal outside its normal habitat or range and therefore unidentified by the observers unfamiliar with the type. The animal in question clearly was a large male

THE ELEPHANT SEAL

The elephant seal, or sea elephant, is the largest of the fin-footed mammals or pinnipeds (the groups of meat-eating water mammals, like seals and walruses, whose four limbs are flippers); it is even larger than the walrus. Males commonly reach 18 feet in length and can weigh up to 5,000 pounds. The male has a flabby snout of about 18 inches that inflates with air when the animal is excited or angry; release of the air produces a deep roar.

elephant seal." He suggested that the creature wandered up through the mouth of the Mississippi River to the White River, which branches off in east-central Arkansas.

Sources:

Mackal, Roy P., *Searching for Hidden Animals: An Inquiry into Zoological Mysteries,* Garden City, New York: Doubleday and Company, 1980.

Folklore in the Flesh

- FLYING DRAGONS AND OTHER SKY SERPENTS

- FAIRIES

- COTTINGLEY FAIRY PHOTOGRAPHS

- MERFOLK

- WEREWOLVES

- THUNDERBIRDS

- THE SEARCH FOR NOAH'S ARK

Folklore in the Flesh

Woodcut from c. 1500 of an armored knight being chased by a flying dragon.

FLYING DRAGONS AND OTHER SKY SERPENTS

From ancient times through the twentieth century, sightings of flying dragons and snakes have been recorded. Such stories first appeared in medieval writings. Henry, Archdeacon of Huntington in medieval England, noted in *Historia Anglorum* that in A.D. 774, "red signs appeared in the sky after sunset, and horrid serpents were seen in Sudsexe, with great amazement." Nineteen years later a similar sight was recorded and construed as a sign of worse things to come: "Terrible portents appeared.... These were exceptional flashes of lightning, and fiery dragons were seen flying in the air, and soon followed a great famine."

According to the publication *Knighton's Continuator,* in April 1388 a "flying dragon was seen ... in many places." On December 5, 1762, a "twisting serpent" lit up the sky as it slowly lowered itself over Bideford, Devonshire, England, then disappeared. It had been visible for six minutes.

Though we expect to read of unbelievable magical events in early writings, like those of the Middle Ages, it can be disturbing to find them recorded in fairly recent times. In the mid-1800s, for example, Nebraska settlers claimed to have witnessed equally strange sights. Western historian Mari Sandoz noted that "back in the hard times of 1857-58 there were stories of a flying serpent that hovered over a Missouri River steamboat slowing for a landing. In the late dusk it was like a great undulating [wavy] serpent, in and out of the lowering clouds, breathing fire, it seemed, with lighted streaks along the sides." A frontier folksong from the period refers to a "flyin' engine/ Without no wing or wheel/ It came a-roarin' in the sky/ With lights along the side/ And scales like a serpent's hide."

The song's mention of a "flyin' engine" may suggest that the object was some sort of unearthly machine (UFO?) rather than a living creature—but other tales seem to demonstrate otherwise. In June 1873 farmers near Bonham, Texas, saw an "enormous serpent" in a cloud and were "seriously frightened." The *Bonham Enterprise* reported: "It seemed to be as large and as long as a telegraph pole, was of a yellow striped color, and seemed to float along without effort. They could see it coil itself up, turn over, and thrust forward its huge head as if striking at something, displaying the maneuvers of a genuine snake."

When the *New York Times* heard of the account a few days later, it called it "the very worst case of delirium tremens on record." The *Times* had nothing to say about another incident, however, that was recorded in a Kansas newspaper, the *Fort Scott Monitor,* a few days earlier. Its June 27 issue reported that at sunrise on the previous morning, "when the disc of the sun was about halfway above the horizon, the form of a huge serpent, apparently perfect in form, was plainly seen encircling it and was visible for some moments."

Since frontier newspapers were full of practical jokes and tall tales, it is hard to take many of these accounts of flying serpents seriously. Consider this story of a man who signed himself "R. B." in a letter to the editor of the *Frederick News,* a Maryland paper, on November 29, 1883. He reported that at 6:30 one morning, while he was standing on

"It seemed to be as large and as long as a telegraph pole, was of a yellow striped color and seemed to float along without effort."

A dragon.

a hilltop, a "monstrous dragon with glaring eye-balls, and mouth wide open displaying a tongue, which hung like a flame of fire from its jaws, reared and plunged" in the sky over Catoctin Mountain.

More Flying Snakes and Other Terrors

In the late 1930s, J. L. B. Smith, a South African chemist with a keen interest in ichthyology (the study of fish), and an associate, Marjorie Courtenay-Latimer, made zoological history when they discovered the coelacanth, a large fish that—to that point—had only been known through fossil records and was thought to have been extinct for some 60 million years.

Smith was also fascinated by reports of other animals generally thought to no longer exist. For a time he exchanged letters with members of a German missionary family; they told him that while living near Mount Kilimanjaro (in northeast Tanzania near the Kenya border), one of them had had a close sighting of a "flying dragon." But the creature was known even before this incident, from the many accounts that native witnesses had reported.

Flying snake.

For her part, Courtenay-Latimer once investigated reports of similar creatures in southern Namibia (then South-West Africa). In one case, native shepherds had walked off their jobs after complaining that their employer, the white owner of a large ranch, did not take their reports of a large flying snake that lived in the nearby mountains seriously. With no one left to watch the livestock, the farmer sent his 16-year-old son to the site. When the boy failed to return that evening, a search party set out to the mountains to look for him. He was found unconscious.

Even after regaining consciousness, the young man could not speak for three days because—his doctor said—he was in shock. Finally, the son related that he had been relaxing beneath a tree when a sudden roaring noise, like a powerful wind current, startled him. As he looked up, he saw a huge "snake" flying down from a ridge. The closer it got, the louder the roaring became. All around the sheep were scattering. The creature landed in a cloud of dust. The boy noticed a strong odor, like burned brass, and at this point passed out.

Courtenay-Latimer arrived on the scene soon afterward. She interviewed witnesses, including other farmers and local police officers, and examined some marks that the creature had reportedly left on the ground. She was told that police had seen the creature disappear into a crack in the mountain; sticks of dynamite were thrown into the opening, which, on combustion, brought a low moaning sound from within—and then silence. The creature was never seen again.

Roy P. Mackal, a scientist who investigates and writes about mysterious animals, contacted Courtenay-Latimer about the account some years later. He wrote, "A snake, even a very large one, hurtling or falling over a ledge or mountain precipice hardly would disturb the air as described. In fact, it is hard to attribute such a disturbance even to a large gliding creature, suggesting instead that some kind of wing action must have been involved." And he wondered, "Could some species of pterodactyl with elongated body and tail still survive?"

In another instance, the *New York Times* reported that on May 27, 1888, in Darlington County, South Carolina, three women strolling through the woods "were suddenly startled by the appearance of a huge serpent moving through the air above them. The serpent was ... sailing through the air with a speed equal to that of a hawk or buzzard but without any visible means of propulsion." According to the account, the frightening creature was around 15 feet long and "was also seen by a number of people in other parts of the county early in the afternoon of the same day." These observers noted that the animal made "a hissing noise which could be distinctly heard."

Sources:

Evans, Jonathan D., "The Dragon," *Mythical and Fabulous Creatures: A Source Book and Research Guide,* edited by Malcolm South, New York: Greenwood Press, 1987.

Sandoz, Mari, *Love Songs to the Plains,* Lincoln, Nebraska: University of Nebraska Press, 1966.

FAIRIES

As he walked down a country road near Barron, Wisconsin, one summer night in 1919, 13-year-old Harry Anderson saw a very strange sight. The bright moonlight revealed 20 little men, walking in single file in his direction! But when they passed, they paid no attention to him. Young Anderson noticed that the men were dressed in leather knee pants held up by suspenders. They wore no shirts, were bald, and their skin was pale white. Though all were making "mumbling" sounds, they did not seem to be communicating with one another. Terrified, Anderson continued on his way and did not look back. The bizarre event remained vivid in his memory for the rest of his life.

To most people in the modern world, fairies are no more than imaginary creatures. They are usually seen in children's entertainment and are almost always portrayed as tiny, winged, and good-hearted. This widespread version of the fairy comes from romantic literature. It does not reflect worldwide fairy folklore, where the beings are described very differently.

Tradition and Its Mysteries

A century or two ago, Anderson would have had little doubt about the identity of the strange figures he had come upon. This would be especially true if he had lived in a Celtic country (Brittany, Ireland, Scotland, Wales), where, according to popular belief, the roads, rocks, caves, fields, rivers, lakes, and forests abounded with creatures of highly unpredictable natures.

Contrary to popular modern images, the fairies of folklore never had wings and were not necessarily kind-hearted.

Only the unwise and unwary called these creatures "fairies," because they did not like to hear their proper name spoken. Because a fairy could be listening at any time, Celtic countryfolk used various flattering names—such as the "good people," the "Gentry," the "honest folk," or the "fair tribe," to describe them.

There was a wide variety of fairy folk. Still, they were more or less human in form, though sometimes taller or shorter (and never bore wings, despite popular modern portrayals). And their behavior was recognizably human. They had governments, societies, occupations, art and music, and conflicts. They married, had children, waged war, and died. At the same time they possessed supernatural powers—which made them bewildering and sometimes dangerous. Few human beings wanted the company of fairies, and most went out of their way to avoid them!

How fairy beliefs began is, according to Stewart Sanderson, "one of the most difficult problems in the study of folklore." Folklorists and anthropologists have theorized that the original fairies were perhaps human members of conquered races who took to the hills to escape their captors and whose descendants were sighted at rare times and mistaken for supernatural beings. It has also been suggested that fairies were the old gods and spirits that man believed in before modern religions took hold. And sometimes, in Christian countries, fairies have been associated with fallen angels. Regardless of origin, belief in fairies has existed in practically every traditional culture. Indeed, the notion that hidden races share the earth with us has been around for most of human history.

An Early Study

Folklorists have learned about fairies through myths, legends, and tales told by countryfolk or found in old printed sources. One of the great early studies on the subject was Robert Kirk's *The Secret Common-Wealth* (1691). Kirk was a Presbyterian clergyman of the Scottish Highlands who had a keen interest in the supernatural lore of the area. He was convinced that fairies were real, for how could such a widespread belief, he asked, "spring of nothing?" For Kirk, fairies were of a "middle nature between man and angel," with bodies "somewhat of the nature of a condensed cloud," and dressed and spoke "like the people and country under which they [lived]." They were usually invisible to the human eye but could sometimes be heard. They often traveled through the air and could steal anything they liked, from food to human babies! People with "second sight" (psychics) were sometimes more able to see them.

Few modern scholars have admitted to believing in fairies. The major exception was W. Y. Evans-Wentz, author of the well-regarded *The Fairy-Faith in Celtic Countries,* first published in 1911. Evans-Wentz, an anthropologist of religion with a Ph.D. from Oxford University, traveled throughout the British Isles and Brittany, on France's northwest coast, and reported the results in a thick book that remains a classic of folklore studies. Evans-Wentz also believed in genies, demons, and other extraordinary beings.

Bad Luck Follows Capture

Years ago, when belief in fairies was still strong, folklorists were able to collect firsthand accounts from fairy witnesses. One such tale was told by an old, blind Irish farmer. The farmer claimed that some years earlier he had captured a fairy, a dark-skinned, two-foot-high figure wearing a red cap, green clothes, and boots.

"I gripped him close in my arms and took him home," the farmer related. "I called to the woman [the farmer's wife] to look at what I had got. 'What doll is it you have there?' she cried. 'A living one,' I said, and put it on the dresser. We feared to lose it; we kept the door locked. It talked and muttered to itself queer words.... It might have been near a fortnight since we had the fairy, when I said to the woman, 'Sure, if we show it in the great city we will be made up [rich].' So we put it in a cage. At night we would leave the cage door open, and we would hear it stirring through the house."

Soon, however, the fairy escaped. Not long afterward the man lost his sight. Other bad luck befell the couple—which the man viewed as the fairy's punishment for its imprisonment.

Midwife to a Troll

Another, earlier fairy episode had a happier ending. It was related in a sworn statement given by Swedish clergyman P. Rahm on April 12, 1671. It seems that late one evening 11 years earlier as he and his wife were sitting and talking in their farmhouse, a little man came to their door. Apparently the fellow's wife was in childbirth, and he begged Mrs. Rahm to go with him and help her. The couple did not know what to do, for they knew that the little man—"of a dark complexion, and dressed in old gray clothes" was a troll. They feared letting Mrs. Rahm go with him, but they also feared the evil that trolls were known to bring upon humans who did not treat them respectfully. After the little man repeated his request several times, Mr. Rahm "read some prayers over [my] wife to bless her, and bid her in God's name go with him."

Mrs. Rahm reported that she seemed to travel on the wind to the little man's dwelling, where his wife lay, in much pain, in a small, dark room. With Mrs. Rahm's help, however, the troll's wife delivered a healthy child. The little man thanked Mrs. Rahm, and she was carried home on the wind as before. The next day, she found pieces of silver on a shelf in the farmhouse sitting room.

Grateful Fairies

Mari Sion of Llanddeusant, Anglesey, Wales, told a folklorist of her own early twentieth-century experience with a fairy family. She recalled that one moonlit night, as she, her husband, and their chil-

With Mrs. Rahm's help, the troll's wife delivered a healthy child. The little man thanked Mrs. Rahm, and she was carried home on the wind.

dren were sitting by the fire, there was a knock on their door. The callers were a man, woman, and baby. The largest of them, the man, was only two feet tall. "I should be thankful for the loan of a bowl with water and a coal of fire," the woman said. "I should like to wash this little child. I do not want them at once. We shall come again after you have gone to bed."

Mrs. Sion put out the requested items before she and her family went to bed. During the night they could hear the comings and goings of the little people. In the morning, the Sions found everything in order, except for the bowl, which lay upside down. Underneath it were four shillings.

Other Sightings

Edward Williams, a respected British clergyman, wrote about a fairy experience he had had in 1757, when he was seven years old. Playing in a field in Wales with some friends, Williams and his companions saw seven or eight tiny couples dressed in red—each figure carrying a white kerchief—some 100 yards away. One of the little men chased the children and nearly caught one, who, according to Williams, got a "full and clear view of his ancient, swarthy, grim complexion" before escaping. During the chase one of the other figures shouted at the pursuer in an unknown language. The incident puzzled Dr. Williams all of his life.

The Rev. Sabine Baring-Gould, a nineteenth-century historian and folklorist, also wrote about his fairy experience as a child. When he was four years old and traveling in a carriage with his parents, "I saw legions of dwarfs of about two feet high running along beside the horses; some sat laughing on the pole, some were scrambling up the harness to get on the backs of the horses." His parents saw nothing. Baring-Gould's wife and son also experienced fairy sightings.

One moonlit night in 1842, a Stowmarket, England, man came upon strange fairy activity. He related the incident to a local historian. While passing through a meadow he spotted a dozen fairies, "the biggest about three feet high, and small ones like dolls. Their dresses sparkled as if with spangles.... They were moving round hand in hand in a ring, no noise came from them. They seemed light and shadowy, not like solid bodies." He could not see their faces because he was some distance away. Fearful that the fairies would discover him, he sped home. When the man brought three companions back to the site to observe the fairies, they were gone.

Over a century later, on April 30, 1973, an educated London woman named Mary Treadgold was traveling by bus through the Highlands of Scotland. When the vehicle pulled over to the side of a narrow road near the town of Mull to let an oncoming car pass by, Treadgold reflexively glanced out the window. There, in a peat field, she saw a "small figure, about 18 inches high, a young man with his foot on a spade ... arrested in the act of digging." She noted that "he had a thin, keen face" and brightly colored clothing. A tiny sack "stood at his side. He was ... not a dwarf, nor a child, nor ... a plastic garden gnome. He was a perfectly formed living being like any of us, only in miniature." When the bus resumed its course, the figure was lost from view. Later, when Treadgold asked a Highland acquaintance about the sighting, the woman related that "friends of hers had seen similar small people on Mull, and that Mull was known for this."

Fairy Music

Many people claimed to have heard fairy music. Isle of Man fiddler William Cain swore he heard music coming from a brightly lit glass palace that he spotted one night in a mountain glen. He stopped and listened, then went home and learned the tune, which he later performed widely. In the summer of 1922, while sitting on the bank of England's Teign River, composer Thomas Wood heard a strange voice calling him by his first name. Though he searched with binoculars, he could not find the speaker. Then he heard, "overhead, faint as a breath," then ever louder, "music in the air. It lasted 20 minutes," he told writer Harold T. Wilkins. "Portable wireless sets [radios] were unknown in 1922.... This music ... sounded like the weaving together of tenuous fairy sounds." Listening with great care, he wrote down the notes.

In 1972, while strolling along the shore of a peninsula in Scotland's Western Highlands, American folksinger Artie Traum heard "thousands of voices" chanting in a strange harmony to the sound of fiddles and pipes.

Today, belief in fairies is all but extinct. That is, except in Iceland, where a recent university survey showed that as much as 55 percent of the population thought the existence of elves (*huldufolk* or "hidden people") certain, probable, or possible. Only 10 percent rejected the idea altogether. Belief is so strong there that construction and road projects are sometimes delayed so that psychics can negotiate with the invisible folk who dwell in Icelandic fields, forests, rocks, and harbors. A 1990 *Wall Street Journal* article noted that "humans and huldufolk usu-

Jacques F. Vallee

(1910-1986)

Jacques Francis Vallee was born in France. After receiving a master's degree in astrophysics, he moved to the United States. While attending Northwestern University, he met J. Allen Hynek, chief scientific consultant on UFOs to the Air Force (also see entry: Unidentified Flying Objects), who headed the astronomy department. Hynek, a one-time UFO doubter, was becoming convinced that there was much to be learned from open-minded UFO research. After Vallee and Hynek began meeting to discuss their views in 1963, Hynek completely broke with the Air Force's anti-UFO line.

Vallee obtained his Ph.D. and ran a computer business, but continued writing and investigating UFOs. In his 1969 book *Passport to Magonia*, he surprised his readers by suggesting that UFO phenomena were beyond science's ability to describe and categorize. To understand UFOs, he said, one would need to immerse oneself in traditional supernatural beliefs in fairies and other fabulous beings.

Vallee believed that most unexplained sightings had their origins in another reality beyond our knowledge. Therefore UFOs and other oddities appear to us in the ways our culture conditions us to expect. Thus, a nineteenth-century Irish peasant saw elves, while his or her modern counterpart might see extraterrestrial humanoids. Vallee wrote many books arguing that UFOs are neither space visitors nor hallucinatory "psychosocial" phenomena; he believed that UFOs are supernatural occurrences.

ally get on well. Midwives have told [folklorist Hallfredur] Eiriksson about delivering elf babies. Farmers say they have milked elf cows. Sometimes, the two peoples fall in love, though affairs of the heart often end badly."

Fairies or Aliens?

It is likely that most fairy sightings would be reported differently today; in this age of UFOs, an encounter with "little people" would no doubt be treated as a meeting with space visitors. In fact, UFO literature contains a handful of incidents that feature elements one might find in traditional fairy lore.

Some writers, in fact, have suggested that UFO sightings and fairy experiences are one and the same. In his *Passport to Magonia* (1969), Jacques Vallee wrote that supernatural shape-shifting beings exist, and that they can appear—depending on the observer's beliefs—as fairies or extraterrestrials. More recent theorists, like Hilary Evans, writing in *Gods, Spirits, Cosmic Guardians* (1987), argue that all encounters with extraordinary beings occur in altered states of consciousness and are hallucinations. Though this psychosocial explanation seems to make some sense, it is actually as weak as the far-fetched ideas of Vallee. Both theories lack physical evidence to support them.

Sources:

Briggs, Katherine, *An Encyclopedia of Fairies: Hobgoblins, Brownies, Bogies, and Other Supernatural Creatures,* New York: Pantheon Books, 1976.
Doyle, Arthur Conan, *The Coming of the Fairies,* New York: George H. Doran Company, 1922.
Evans, Hilary, *Gods, Spirits, Cosmic Guardians: A Comparative Study of the Encounter Experience,* Wellingborough, Northamptonshire, England: The Aquarian Press, 1987.
Vallee, Jacques, *Passport to Magonia: From Folklore to Flying Saucers,* Chicago: Henry Regnery Company, 1969.

COTTINGLEY FAIRY PHOTOGRAPHS

In 1917 two young English girls, Frances Griffiths, 10, and her 13-year-old cousin Elsie Wright, were living in Cottingley, near Bradford, Yorkshire. Frances's father was a soldier fighting in World War I; she and her mother were staying at the Wright house until he returned home. One day Frances entered the house soaking wet and claimed that she had fallen into the brook while playing with the

Frances Griffiths and the Cottingley fairies.

When Elsie's doubting father developed the picture he saw an image of Frances facing the camera as four miniature winged women dressed in filmy clothing danced in front of her!

fairies the girls had befriended in a nearby glen. Her mother did not believe her, and Frances was punished.

Feeling sorry for her cousin and best friend, Elsie hit upon an idea: they would borrow her father's camera and photograph the fairies.

Elsie told her father that she was going to take a picture of her cousin, and he gave her his camera and a single glass plate (which was used in photography before the advent of film). An hour later the girls returned and said that they now had proof that the fairies were real. When Elsie's doubting father developed the picture he saw an image of Frances facing the camera as four miniature winged women dressed in filmy clothing danced in front of her!

The girls refused to admit that they had photographed paper cutouts, even though their parents were certain that they had. Still, one month later, Mr. Wright gave the girls the camera and another plate. They returned with a second picture, this time showing a sitting Elsie signaling an elflike figure to jump up on her lap! Convinced that the joke was getting out of hand, Wright barred the girls from using the camera again.

Before the war Frances had lived in South Africa. When she wrote a friend there she enclosed copies of the two fairy photographs. On the back of one she noted, "Elsie and I are very freindly [sic] with the beck [brook] fairies. It is funny I did not see them in Africa. It must be to [sic] hot for them there." When rediscovered and published (in the *Cape Town Argus,* November 25, 1922), Frances's words would be taken as evidence of the girls' sincerity and the photographs' authenticity.

The Case

The photographs would become the subject of one the strangest and most hotly debated cases in the history of photography. When Polly Wright, Elsie's mother, attended a lecture on folklore in 1920, she mentioned the photographs to the speaker. He requested the prints, which were shown to H. Snelling, an expert on photography. Snelling's pronouncement that the pictures were genuine would be quoted for decades afterward. (It would not be known until 1983 that he had retouched the first photograph—badly over-exposed—transforming it into the clear version that was widely seen.)

A well-known London figure, Edward L. Gardner, befriended the Wrights. At the urging of Sherlock Holmes author Sir Arthur Conan Doyle, Gardner took the photos to the Kodak laboratory in London. There, Doyle reported, "two experts were unable to find any flaw, but refused to testify to the genuineness of them, in view of some possible trap." When Gardner gave Elsie a modern camera, she and Frances provided three more fairy photographs!

In December 1920 *The Strand* magazine published Doyle's article on the first two pictures, and the next March a follow-up piece included the later three. The story received worldwide publicity. Much of it was unfavorable, though, focusing on how the illustrious author could have fallen for such an obvious hoax.

Cottingley photograph of Elsie Wright and fairy.

Yet attempts to discredit the photographs were not successful. Models for the Cottingley fairy figures could not be found. Furthermore, when spiritualist Geoffrey Hodson visited the beck in the girls' company, he reported seeing many such beings.

There would be no more Cottingley fairy photographs after that. Still, the mystery lived on. Doyle wrote an entire book on the case, *The Coming of the Fairies,* in 1922. In 1945 Gardner published a book-length account of the episode, and the photographs were reprinted in newspapers and magazines from time to time. Elsie and Frances seemed to stand by the pictures, but gave vague answers when asked about them. Then, in 1972, Elsie sent two cameras, along with other materials related to the case, to Sotheby's (an auction house) for sale; with them she included a letter confessing for the first time that the photographs were fakes. Sotheby's returned the letter, failing to recognize what it had in its possession.

In the early 1980s the *British Journal of Photography* began a reinvestigation of the case, based on research by editor Geoffrey Crawley. Frances and Elsie gave him their signed, formal confessions in early 1983. The two had agreed that the truth—that the pictures were a "practical joke" that "fell flat on its face"—would be withheld until the deaths of the photographs' major supporters, Doyle, Gardner, and Gardner's son Leslie.

A gifted young artist, Elsie had created the figures using fairies pictured in a popular children's book, *Princess Mary's Gift Book,* as her models. But to the end, the two women would not reveal the photographic techniques they used, promising to reveal them in books they were writing. Both died, however, before finishing the works. Nonetheless, Frances would always insist that while the photographs were fake, she *had* seen real fairies in the beck!

Sources:

Doyle, Arthur Conan, *The Coming of the Fairies,* New York: George H. Doran Company, 1922.
Gardner, Edward L., *Fairies: The Cottingley Photographs and Their Sequel,* London: Theosophical Publishing House, 1945.
Vallee, Jacques, *Passport to Magonia: From Folklore to Flying Saucers,* Chicago: Henry Regnery Company, 1969.

MERFOLK

Belief in merfolk—mermaids and mermen—has been around since ancient times. The Babylonian god Oannes, who rose from the Erythraean Sea to grant knowledge and culture to the human race, was said to be human to the waist and fish-shaped from the waist down. Merfolk-like gods and goddesses were also worshiped in Syria, India, China, Greece, and Rome. In later centuries a worldwide folklore would develop around these creatures and actual sightings would be claimed.

> **THE GREY SELKIE OF SULE SKERRIE, A CLASSIC FOLK BALLAD**
>
> I am a man upon the land
> I am a selkie [seal]
> in the sea.

One of the earliest people to write about the creatures was first-century Roman naturalist Pliny the Elder. He had no doubt that Merfolk existed because of the many sightings reported by coastal residents. He even noted, "Many of these Nereides or Mermaids were seen cast upon the sands, and lying dead."

In northern Europe the merfolk legend took on a different twist; seal-folk or selkies were seals when they lived in the water. But when they wished to pass themselves off as people on land, they simply removed their seal skins! In many folktales merfolk do the same. They shed their fishy forms, marry land-bound mortals, and

For a sailor, seeing a mermaid was often a sign of coming death at sea.

even have children. Then an overwhelming homesickness for the sea overtakes them, and they are gone in a splash.

Sailors usually interpreted the sighting of a mermaid as a sign of coming death, often in the fierce storm that would frequently follow such a visitation. In the traditional ballad "The Mermaid," a ship's crew spots one of the creatures sitting on a rock with a comb and a glass in its hand. The captain states: "This fishy mermaid has warned me of our doom/ And we shall sink to the bottom of the sea ..../ And three times around spun our gallant ship .../ And she went to the bottom of the sea."

Merfolk of Scotland

But mermaids were more than legendary creatures. Actual sightings by reliable witnesses were reported throughout the Middle Ages. And such sightings continued into modern times.

On January 12, 1809, two women standing on a beach at Sandside, Caithness, in remote northeastern Scotland, saw what looked like the face of a young woman—"round and plump and of a bright pink hue"— in the sea. The creature then disappeared into the water, reappearing a short time later. When the women were able to observe more of its body, they could see that it had well-formed human breasts. From time to time it lifted a long, thin white arm above the waves and tossed back its long green hair.

After one of the witnesses published her account of the sighting, William Munro wrote a letter to the *London Times* (September 8, 1809) recalling his own mermaid experience. Twelve years earlier, while walking along the shore of Sandside Bay, he spotted what looked like "an unclothed female, sitting upon a rock extending into the sea, and apparently in the action of combing its hair, which flowed around its shoulders, and was of a light brown color."

Munro reported that the creature's "forehead was round, the face plump, the cheeks ruddy, the eyes blue, the mouth and lips of a natural form," and that "the breasts and the abdomen, the arms and fingers [were] of the size of a full-grown body of the human species." The creature did not seem to notice him as he watched, and it continued to comb its hair, "which was long and thick, and of which it appeared to be proud." Then, after a few minutes, the mermaid slipped back into the sea.

It seemed that such creatures were particularly active off the coast of Scotland during this period. In a long survey of mermaid and merman sightings, the *London Mirror* of November 16, 1822, listed a similar account officially sworn to by young John McIsaac. On the afternoon of October 13, 1811, the fellow saw a strange creature "on a black rock on the seacoast." He stated that "the upper half of it was white, and of the shape of a human body" (though its arms were unusually short), while its bottom half was scaly, shiny, and ranged from reddish-grey to reddish-green in color. "The animal was between four and five feet long" and had a tail that it spread "like a fan."

As with Munro's mermaid, the creature McIsaac observed had long hair that it liked to stroke. After two hours of lying on the rock, McIsaac's creature "tumbled clumsily into the sea," allowing him to see "every feature of its face, having all the appearance of a human being." Because the creature was now half covered with water and was "constantly, with both hands stroking and washing its breasts," McIsaac could not tell if it was a female. The animal eventually disappeared.

Five days later another eyewitness gave sworn testimony before the same local sheriff as had McIsaac. Katherine Loynachan stated that on the afternoon of October 13, the same time day on which McIsaac had spotted his mermaid, as she was herding cattle near the seashore, she saw a creature slide off one of the rocks and drop into the water, surfacing six yards out. It had long, dark hair, white skin on its upper body, and dark brown skin on its lower half, which was fishlike. When it reapproached the shore she saw its face clearly and it looked small and white—like a child's. And, as noted in other sightings, it was "constantly rubbing or washing its breast." After a time it swam away.

SEAWEED GREEN

Unlike the golden-haired mermaids of legend and Disney movies, those in eyewitness accounts usually have darker hair, from green to black. Many witnesses of mermaids noted their long, thick hair, of which—green or not—they seemed proud.

At first Katherine did not trust what she saw; she told herself that a boy had fallen out of a boat and was seeking rescue. As her father later recalled, she came running home to tell him about a strange boy who was swimming along the shore. She and her father and mother then went looking for the lad but found nothing.

A series of sightings took place off Scotland's west coast in the summer of 1814. One incident involved a group of children, who saw what they at first thought was a drowning woman. But, according to a letter published in the *York Chronicle* (September 1), closer examination revealed what appeared to be a mermaid: the upper half resembled a fair, long-haired, rosy-cheeked woman (except that its lower arms and hands were as small as a child's), and the bottom half looked like "an immense large cuddy fish ... in color and shape." Some of the children brought nearby farmers to the scene, and one of them prepared to shoot the creature with his rifle. But the others kept him from doing so. Instead he whistled at the mermaid, which glanced at him. It "remained in sight for two hours, at times making a hissing noise like a goose," the *Chronicle* reported. The creature was seen two more times, "always early in the morning and when the sea was calm."

On August 15 of that year, two fishermen were a quarter-mile from shore at Port Gordon when they spotted a merman. According to an account in the *Caledonian Mercury,* it had a dark face, small eyes, flat nose, large mouth, and very long arms. Soon afterward it was joined by what the men guessed was its mate, for this second creature had long hair, fair skin, and breasts. Frightened by the strange sight, the fishermen raced to shore, with the creatures gazing at them all the while!

Around 1830 people working along the shore of Benbecula island, off Scotland's northwest coast, spotted a small creature, half woman and half fish, turning somersaults in the water. Some men tried to capture it, but they had no success. Finally, a boy hit it on the back with some stones, and it disappeared. A few days later the body washed up on shore two miles away.

District Sheriff Duncan Shaw examined the body carefully. He reported that "the upper part of the creature was about the size of a well-fed child of three or four years of age, with an abnormally developed breast. The hair was long, dark, and glossy, while the skin was white, soft and tender. The lower part of the body was like a salmon, but without scales." The creature was buried in the presence of a number of island residents at a graveyard in Nunton. "The grave is pointed out to this day," folklorist R. MacDonald Robertson stated in 1961. "I have seen it myself."

DUGONGS AND MANATEES

Dugongs and manatees are large aquatic plant-eating mammals from the sirenian, or sea cow, family. They have thick, heavy bodies with weak front flippers, no hind legs, and tails ending in flattened fins. Their gray skin is hairless, except for whiskers on the face. The female has a pair of mammary glands (breasts) on her chest and holds her pup in her flippers while nursing. It has been suggested that the manatee, which surfaces to nurse, is the source of mermaid sightings. While both animals are most often found in warm, shallow, protected waters, the dugong may be sighted farther out to sea.

Merfolk in the Americas

Christopher Columbus is easily the most famous observer of mermaids. On his voyage of discovery in the West Indies, he saw three of them "leaping a good distance out of the sea" and found them "not so fair as they are painted." From the behavior he described, it is more likely that he had spotted a trio of sea mammals known as dugongs.

Other sightings in the Americas followed. While explorer John Smith was sailing through the West Indies in 1614, he spotted a young woman in the water. So attractive was she that Smith began to "feel the first pains of love." That is, until he discovered that "from below the waist the woman gave way to the fish!" Four years earlier, while sailing a small boat into a harbor at St. John's, Newfoundland, Canada, a Captain Whitbourne saw a strange creature that resembled a woman swimming in his direction. Alarmed, he quickly backed away. The creature then turned around and tried to board another boat, this one belonging to William Hawkridge. He banged it on the head! It disappeared under the water.

That mermaid was lucky compared to a merman sighted during the seventeenth century in Casco Bay, off the southern coast of Maine. When it tried to enter the boat of a Mr. Mitter, the boat owner reportedly slashed off one of its arms. The creature sank, "dying the waters purple with its blood," according to one writer. Not long after that, in the waters off Nova Scotia, Canada, crews of three French vessels sighted another merman. They chased it and tried to capture it with ropes but without success. "He brushed

Folklore in the Flesh

his mossy hair out of his eyes which seemed to cover his body as well," the captain of one of the ships recorded.

Another sighting was recorded by a very reliable witness, New World explorer Henry Hudson, for whom the Hudson River is named. On the evening of June 15, 1610, two members of his crew observed a mermaid. She had white skin, long black hair, and "her back and breasts were like a woman's"; she also had "the tail of a porpoise." The men got a good look at her because she came "close to the ship's side, looking earnestly on the men." Commenting on the sighting, nineteenth-century naturalist Philip Gosse remarked: "Seals and walruses must have been as familiar to those polar mariners as cows to a milkmaid. Unless the whole story was a concocted lie between the two men, reasonless and objectless, and the worthy old navigator doubtless knew the character of his men, they must have seen some form of being as yet unrecognized."

In 1797 a Dr. Chisholm visited the tiny island of Berbice in the Caribbean. There Governor Van Battenburgh and others told him of repeated sightings in the island's rivers of strange creatures that the Indians call *mene mamma* (mother of waters). In his 1801 book *Malignant Fever in the West Indies,* Chisholm wrote: "The upper portion resembles the human figure.... The lower portion resembles the tail

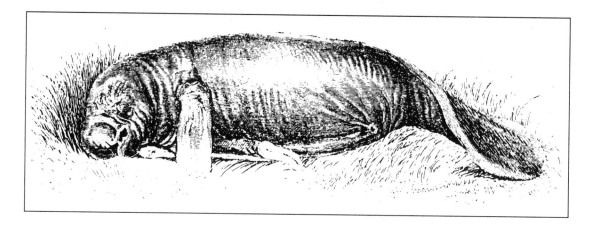

A sea cow.

portion of a fish ... not unlike that of the dolphin.... They have been generally observed in a sitting posture in the water, none of the lower extremity being discovered until they are disturbed.... They have been always seen ... smoothing their hair, or stroking their faces and breasts with their hands, or something resembling hands.... They have been frequently taken for Indian women bathing.

Theories

One proposed explanation for mermaid sightings is that they are sea cows—manatees and dugongs—which, in the words of scientist Richard Carrington, "became 'transformed' into a mermaid by the expectant attention of the superstitious mariners. However, a survey conducted by Gwen Benwell and Arthur Waugh, authors of *Sea Enchantress* (1965), shows that nearly three-quarters of such sightings occurred far from areas where dugongs and manatees are known to exist. Secondly, and more importantly, the animals hardly resemble the creatures described in the sightings.

But the sea cow explanation should not be dismissed in all cases. The people of New Ireland, an island province of Papua New Guinea, for instance, frequently reported seeing what sounded like merfolk: creatures that looked human down to their waists and had legless lower trunks ending in two side fins. They called the creatures **ri**, and when anthropologist Roy Wagner visited the island in the late 1970s, they told him the animals resembled the mermaids on tunafish cans. Understandably, he was intrigued. After experiencing a sighting himself, Wagner was positive that the creatures were not dugongs.

But a February 1985 expedition by American scientists produced underwater photographs of a ri—and it was without a doubt a dugong. One part of the mystery was solved. But expedition member Thomas R. Williams still wondered "how myths of merfolk can arise and persist in the face of the obvious reality of the dugong."

Two *Nature* writers proposed a second explanation for merfolk sightings. Studying Norse merman reports, they concluded that atmospheric changes or inversions could create strange optical effects, resulting in distortions on the ocean surface. Thus killer whales, walruses, and even jutting rocks could be perceived by sailors as merfolk. These atmospheric inversions were also responsible for the storms that so often followed merfolk sightings. After reviewing the study, behavioral scientist David J. Hufford felt that the explanation had some merit.

Michel Meurger, a French folklorist and expert on the lore of fabulous water beasts, feels that biological explanations of merfolk sightings are useless. He considers sightings as "visionary experiences," or vivid hallucinations that take their shape from popular superstitions.

Another theory purports that merfolk are simply an undiscovered species. Bernard Heuvelmans, the founder of cryptozoology (the science of unknown animals), stated in a 1986 paper, "Only a still-unrecorded species of recent Sirenia [sea cows], or possibly—though much less likely—an unknown form of primate adapted to sea-life, could explain the abundance and persistence of merfolk reports in certain seas up to modern times." Benwell and Waugh came to the same conclusion. Many dismiss this explanation, however, because no remains of the creatures—often spotted close to shore—have ever been studied scientifically.

Sources:

Beck, Horace, *Folklore and the Sea,* Middletown, Connecticut: The Marine Historical Association/Wesleyan University Press, 1973.

Benwell, Gwen, and Arthur Waugh, *Sea Enchantress: The Tale of the Mermaid and Her Kin,* New York: The Citadel Press, 1965.

Berman, Ruth, "Mermaids," *Mythical and Fabulous Creatures: A Source Book and Research Guide,* edited by Malcolm South, Connecticut: Greenwood Press, 1987.

Costello, Peter, *The Magic Zoo: The Natural History of Fabulous Animals,* New York: St. Martin's Press, 1979.

WEREWOLVES

Human beings have believed in *lycanthropy,* the transformation of a man or woman into a wolf or wolflike human, since ancient times. One of the earliest tales of such a transformation is in Greek mythology: Zeus punished Lykaon for serving him and other gods human flesh by turning him into a wolf. Inspired by this myth, a cult in long-ago Arcadia (Greece) required that each new member sacrifice a human being, which made the sacrificer a "wolf" for nine years. Several references in ancient writings indicate this early fascination with man-wolves.

The word "werewolf" first saw print in the eleventh century. The first half of the name comes from the Teutonic *wer,* meaning "man"; thus a werewolf is a man-wolf. Though known to most of us simply as a subject of horror movies and novels, the werewolf was once feared as a real-life terror.

History has demonstrated that belief in human-animal transformations featured not only wolves, but also bears, big cats, hyenas, and other fierce creatures. But of all of these the werewolf is best known, no doubt because the wolf was the predator most feared by Europeans. Medieval and later accounts tell of attacks by wolves on human beings (also see entry: **Beast of Gevaudan**), usually during wars and hard winters.

Though zoologists today assure us that wolves are generally harmless to people (the harm, in fact, is more frequently done *by* people *to* wolves), "it is difficult," in the words of folklorists W. M. S. Russell and Clair Russell, "to believe that all the past accounts [of wolf attack] are legendary." For "modern wolves have had many generations' experience of fire-arms, and are likely to be more cautious than their ancestors." In fifteenth- to eighteenth-century books on hunting, in fact, "werewolf" was the name given to a wolf that had developed a taste for human flesh.

In northern Europe, wolfmen or "berserkers" were warriors dressed in clothing made of wolf skin. They were well-known murderers and deeply feared. At the same time, however, in Germany it was believed that after death, honored ancestors had become wolves. In the Baltic and Slavic regions of Europe, people worshipped a temperamental wolf god: it could protect, but it could also turn on its faithful without warning. As Christianity rose to power, priests condemned such pagan beliefs as satanic.

At times in the past, werewolves have been considered the devil's agents (also see entry: **Black Dogs**). Religious writers debated

A red wolf.

whether Satan really turned humans into wolves, or whether they were simply perceived as such by those whom the devil had under his spell. Many eventually concluded that only God—whose powers were greater than Satan's—could actually effect such physical transformations.

And some people believed that they themselves were werewolves. Several of these reported that they rubbed themselves with a salve to bring on the transformation. The salve contained hallucination-producing plants like henbane and deadly nightshade. The potion was called "witches' salve," and was also supposedly used by witches to cover themselves before flying off to sabbats—midnight ceremonies of devil worship. By the time of the Renaissance witch trials, many writers felt that both sabbats and human-wolf transformations occurred only in the drugged imagination.

Others writers thought that mental illness was behind a person's belief that he or she could turn into a wolf or had seen others do so. And it was thought that sometimes the devil further confused these already disturbed individuals. In Germany in 1589, for example, Stubbe Peeter was tried for 25 years of hideous crimes, including murder of adults and children (including his own son), cannibalism, incest, and attacks on animals. Peeter claimed to have made a pact with Satan, who provided him with a belt that turned him into a wolf. Nine years later French officials arrested beggar Jacques Roulet after they found him crouching in a bush and covered with blood from the badly mutilated body of a 15-year-old boy discovered nearby. In his confession, Roulet said he had slain the youth while a werewolf, the transformation occurring after he had rubbed himself with a salve.

BIG FOOT MIX

ESSENCE OF THUNDER BIRD

WERE WOLF JUICE

EU DE MOTH MAN

YETI NECTA[R]

Several people reported that they rubbed themselves with a salve in order to become a werewolf.

Modern psychiatry views lycanthropy as a serious mental illness. According to psychiatrists Frida G. Surawicz and Richard Banta, it can be triggered by drug abuse, brain damage, or other causes. Psychoanalyst Nandor Fodor felt the belief in wolf-man transformations "cannot be traced to a point in historic time or to particular civilizations"; rather, it springs from deep within the human mind. Similarly, psychologist Robert Eisler theorized that the wolf, representing nature at its animal fiercest, lies deep in the human subconscious, a kind of memory from a time when early human beings were hunter-killers. In rare cases this buried information may rise up to overwhelm a person's consciousness, forcing him or her to actually identify with the wolf.

Sightings

Werewolves are found not only in mythology, folklore, and popular culture, but in modern sighting reports as well, though these are few and most are poorly documented.

In a 1960 issue of the magazine *Fate,* Mrs. Delburt Gregg of Greggton, Texas, told of an experience with a shape-changing creature. Such reports are very rare in modern times. The other sightings discussed below are simply of creatures that resembled man-wolves; nobody on record claims to have seen one becoming another. Gregg did not make such a claim either, but she came close in a tale that sounds more like the beginning of a werewolf novel than a real-life experience.

One night in July of 1958, while her husband was away on business, Mrs. Gregg moved her bed close to a screened window hoping to catch a cool breeze from an approaching thunderstorm. She dozed off for a short time before she heard a scratching sound on the screen. In a flash of lightning she saw a "huge, shaggy, wolflike creature ... clawing at the screen and glaring at me with baleful, glowing, slitted eyes. I could see its bared white fangs."

She leaped from bed and grabbed a flashlight as the creature fled through the yard and into a clump of bushes. "I watched for the animal to come out of the bushes," she wrote, "but, after a short time, instead of a great shaggy wolf running out, the figure of an extremely tall man suddenly parted the thick foliage and walked hurriedly down the road, disappearing into the darkness."

More common modern American werewolf tales take the form of those told by a number of Ohio residents between July and October 1972. They reported seeing a six- to eight-foot-tall creature, which one witness described as "human, with an oversized, wolflike head, and an elongated nose." Another said it had "huge, hairy feet, fangs, and it ran from side to side, like a caveman in the movies." It also had glowing red eyes. One early morning it reportedly sneaked up behind a trainman working along the tracks in downtown Defiance and whacked him with a piece of lumber.

One day in January 1970, four youths from Gallup, New Mexico, reportedly encountered what they called a "werewolf" along the side

In 1972 some Ohio residents reported seeing a six– to eight–foot–tall creature, which one witness described as "human, with an oversized, wolflike head, and an elongated nose."

of a road near Whitewater. It managed to keep up with their car, which was going around 45 miles per hour! One witness related: "It was about five feet seven, and I was surprised it could go so fast. At first I thought my friends were playing a joke on me, but when I found out they weren't, I was scared! We rolled up the windows real fast and locked the doors of the car. I started driving faster, about 60, but it was hard because that highway has a lot of sharp turns. Someone finally got a gun out and shot it. I know it got hit and it fell down, but there was no blood. I know it couldn't be a person because people cannot move that fast." (This creature, in fact, fits the description of a "skin-walker" —what Navajos of the Southwest call their version of a werewolf—which travels incredibly fast.)

In the fall of 1973, western Pennsylvania experienced a flurry of sightings of strange creatures, some linked to UFO reports. Investigator Stan Gordon noted that one type of being observed "was said to be between five and six feet tall. It was described as looking just like an extremely muscular man with a covering of thick dark hair.... This creature appeared to have superior agility.... From footprints discovered, the stride of the creatures varies between 52 and 57 inches." Two sketches published in *Flying Saucer Review* (July 1974) indicated that the beast strongly resembled traditional werewolves, a fact of which—incredibly—neither witnesses nor investigators took note, however.

REEL LIFE

I Was a Teenage Werewolf, 1957.

A troubled young man—played by a very young Michael Landon in his first feature film appearance—suffers from teen angst and goes to a psychiatrist. The good doctor turns out to be a bad hypnotist, and Landon's regression therapy takes him beyond childhood into his primal past, where he sprouts facial and knuckle hair.

Werewolf of London, 1935.

A scientist searching for a rare Tibetan flower is attacked by a werewolf. He scoffs at the legend, but once he's back in London, he goes on a murderous rampage every time the moon is full. Dated but worth watching as the first werewolf movie made.

The Wolf Man, 1941.

Fun, absorbing classic horror with Lon Chaney, Jr., as a man bitten by werewolf Bela Lugosi. His dad thinks he's gone nuts, and his screaming girlfriend just doesn't understand. Chilling and thrilling!

A Few Explanations

If these werewolf stories are not outright hoaxes, they would seem to point to the presence of extraordinary, otherworldly creatures. Still, more conventional explanations for what witnesses have observed do

exist. For example, in Kansas during July 1974, several people reported coming upon what one newspaper described as a "young child about 10 or 12 years old, with bloody, matted hair, dressed in tattered clothing, running through vines and brush in a wooded area in the northwest edge of Delphos." She was dubbed the "wolf girl." Though local officials never located her, she may well have been a runaway or abandoned child.

Some in the medical community have suggested that sightings of werewolves have really been of individuals with a rare genetic disease called porphyria. Porphyria sufferers are plagued by tissue destruction in the face and fingers, open sores, and extreme sensitivity to light. Their facial skin may take on a brownish cast, and they may also suffer from mental illness. The inability to tolerate light, plus shame stem-

Lon Chaney, Jr., as the Wolf Man in the 1941 film classic.

ming from physical deformities, may lead the afflicted to venture out only at night. "These features," British neurologist L. Illis wrote in a 1964 issue of *Proceedings of the Royal Society of Medicine,* "fit well with the description, in older literature, of werewolves."

Sources:

Cheilik, Michael, "The Werewolf," *Mythical and Fabulous Creatures: A Source Book and Research Guide,* edited by Malcolm South, Connecticut: Greenwood Press, 1987.

Clark, Jerome, and Loren Coleman, *Creatures of the Outer Edge,* New York: Warner Books, 1978.

Fodor, Nancy, *The Haunted Mind: A Psychoanalyst Looks at the Supernatural,* New York: Garrett Publications, 1959.

THUNDERBIRDS

"Its wingspread appeared to be as wide as the streambed, which I would say was about 75 feet."

Many Native North American tribes once believed in "thunderbirds," giant supernatural flying creatures that caused thunder by flapping their wings and lightning by closing their eyes. Thunderbirds were said to war with other supernatural creatures, and they sometimes granted favors to human beings. The mythological beasts can often be seen on totem poles, pillars carved and painted with symbols and mythical or historical incidents.

Thunderbirds in Pennsylvania

In modern times, those who study accounts of strange animals sometimes use the term *thunderbirds* to describe the unlikely giant birds that are seen and reported from time to time. The heavy forests of north-central Pennsylvania have long seemed a favorite spot of the huge creatures; in 1973 Pennsylvania writer Robert R. Lyman declared that "their present home is in the southern edge of the Black Forest.... All reports for the past 20 years have come from that area." He insisted, "Thunderbirds are not a thing of the past. They are with us today, but few will believe it except those who see them."

Lyman himself claimed to have seen one of the birds in the early 1940s. When first observed, it was sitting on a road near Coudersport. It then rose a few feet into the air and spread its wings, which measured at least 20 feet! It flew into the dense woods that lined the high-

way. Like most other witnesses, Lyman thought that the bird looked like a "very large vulture," brown, with a short neck and "very narrow" wings. (Vultures are large birds of prey, related to the hawks, eagles, and falcons, but with weaker claws and a bald head. Their diet is usually made up of flesh from animals they find that are already dead.)

Lyman felt that the specimen he saw was just a young bird. In 1969 the wife of Clinton County sheriff John Boyle saw a huge gray bird land in the middle of Little Pine Creek while she was sitting in front of the couple's wilderness cabin. A few moments later it rose to fly away and "its wingspread," she said, "appeared to be as wide as the streambed, which I would say was about 75 feet." Also that summer three men claimed to have seen a thunderbird snatch up a 15-pound fawn near Kettle Creek.

Just east of Clinton County, over in Jersey Shore, Pennsylvania, many accounts of thunderbirds have been reported over the years. On

October 28, 1970, for instance, several people driving west of town saw a startling sight. One of the eyewitnesses, Judith Dingler, described it as a "gigantic winged creature soaring towards Jersey Shore. It was dark colored, and its wingspread was almost like [that of] an airplane."

Pennsylvania's thunderbird stories have been traced well back into the nineteenth century. Records of the sightings, however, have been spotty. If in print at all, they usually appear as short reports in local newspapers.

CONDORS

Condors are large vultures found in the high peaks of the Andes Mountains of South America and the Coast Range of southern California. They are the largest living birds. They are constant eaters and prefer the remains of already dead animals because they have weak talons and lack the strength of other birds that hunt for their food. They will attack live animals when they must but look for those that are most helpless. Condors have keen sight and are skillful soarers, riding the updrafts around their mountain homes. They are usually seen alone. The Andean condor has black feathers with white wing patches and a white ring of downy feathers around a bare neck; the gray head is also bare. The rare California condor is all black with white wing bands.

Still, sightings of thunderbirds, especially those describing giant vultures or eagles, can be found all across the country. And the reports are so similar that Mark A. Hall, the leading expert on the subject, has pieced together a general description of the birds, based on eyewitness accounts. He notes that what makes them most remarkable are their size and lifting strength—far greater than "those of any known bird living today anywhere in the world." While the dimensions of wingspans, of course, can only be guessed, sometimes measurable objects close by have offered reliable comparisons. Wingspreads seem to range from 15 to 20 feet; the birds themselves appear to be from four to eight feet tall. They are generally dark in color: brown, gray, or black.

Attack of the Giant Vultures

In July 1925, two visitors to the Canadian Rocky Mountains of Alberta spotted what they thought was an eagle high in the sky. As it approached a 7,500-foot mountain peak, they noticed that it was huge and brown and—even more surprisingly—carrying a large animal in its talons (claws). Shouts from the observers caused it to drop the animal, which turned out to be a 15-pound mule-deer fawn.

Bird experts insist that such a report—and there are many more like it—describes the impossible. The largest predatory birds, such as the eagle, attack only "small mammals, reptiles, fish, and perhaps, some other birds," maintains wildlife expert Roger A. Caras, for example. The largest American bird, the rare and endangered California condor, has a wingspan of slightly over ten feet (though one specimen captured early in the century did measure 11' 4"). Even so, its weak talons do not permit it to carry prey; instead, it usually feeds on the remains of animals that are already dead.

Following are some sightings of the vulture variety of thunderbirds:

Kentucky, 1870

A "monster bird, something like [a] condor," landed on a barn owned by James Pepples of Stanford. Pepples fired at the creature, wounding and capturing it. A press account at the time reported that the bird measured "seven feet from tip to tip" and "was of a black color." It is not known what became of the animal.

Illinois-Missouri border, 1948

A number of people reported seeing a huge bird that resembled a condor. And they also claimed that it was about the size of a Piper Cub airplane!

California condor.

Puerto Rico, 1975

During a flurry of unexplained night killings of farm animals and pets, livestock owners sometimes reported being awakened by a "loud screech" and the sound of giant wings flapping. Several witnesses claimed daylight sightings of what one called a "whitish-colored gigantic condor or vulture."

Northern California, October 1975

Residents of a Walnut Creek neighborhood saw a huge bird, over five feet tall, with a "head like a vulture" and gray wings. Five minutes later it flew away, revealing a 15-foot wingspan. Around the same time, a number of people observed the same or a similar bird sitting on a rooftop in nearby East Bay.

Birds Attacking Humans

A remarkable series of events that took place in 1977 attracted wide public attention. They began on the evening of July 25 in Lawndale in central Illinois. Three boys, one of them ten-year-old Marlon Lowe, were playing outside when they saw two large birds in the sky. The animals swooped down towards one of the boys, who jumped into a swimming pool to escape.

They then turned to Marlon, grabbing him by the straps of his sleeveless shirt and lifting him two feet above the ground. As Marlon screamed, his parents, Jake and Ruth Lowe, and two friends, Jim and Betty Daniels, heard him and witnessed the awful sight of the boy held in the talons of a flying bird. Marlon was beating at the creature with his fists, and finally, after carrying him for about 40 feet, it dropped him. By this time Mrs. Lowe was following close behind. With her son safe on the ground at last, she noted that "the birds just cleared the top of the camper, went beneath some telephone wires and flapped their wings—very gracefully—one more time," before they flew off toward some tall trees edging a nearby creek.

According to witnesses, the birds were black, with white rings on their long necks. They had curved beaks and eight- to ten-foot wing-spreads. After checking in books at the library, the Lowes decided that the birds looked like condors.

Bird experts and other officials wasted no time in judging the event impossible—and declaring everyone involved in it liars. The Lowes found themselves the focus of many cruel comments. Marlon himself suffered from nightmares for weeks afterward, though he received no physical injuries in the ordeal. The Lowes and their friends were not the only people, however, who reported seeing strange birds in the area. Other people reported sighting unusually large birds for two weeks thereafter.

Monster Eagles

In his 1975 book *Dangerous to Man,* Roger Caras wrote that while totally untrue, "the stories about eagles carrying off human babies, and even small children, are absolutely endless." Some who have investigated such reports, however, might disagree with Caras, though the ability of a normal-sized eagle—which never weighs much over seven pounds—to carry anything but the smallest animals has never been proven. Yet at least one such kidnapping, however "impossible," has been well documented.

On June 5, 1932, Svanhild Hansen, a 42-pound, five-year-old girl, was taken from her parents' farm in Leka, Norway, by a huge eagle. The bird carried her more than a mile before it dropped her on a high ledge,

> Marlon was beating at the creature with his fists, and finally, after carrying him for about 40 feet, it dropped him.

continuing to circle overhead. When rescuers reached the ledge the child was asleep. Except for a few small scratches, she was unharmed. Zoologist Hartvig Huit-feldt-Kaas spent a month investigating the story and found it "completely reliable." The eagle—if that is what it was—was seen several more times.

There have been many other cases of kidnappings by eagles, though most are considerably less well documented. All of them do not have happy endings. Felix A. Pouchet's 1868 nature encyclopedia, *The Universe*, tells a horrifying story from the French Alps that reportedly took place in 1838:

> A little girl, five years old, called Marie Delex, was playing with one of her companions on a mossy slope of the mountain, when all at once an eagle swooped down upon her and carried her away in spite of the cries and presence of her young friend. Some peasants, hearing the screams, hastened to the spot but sought in vain for the child, for they found nothing but one of her shoes on the edge of a precipice. The child was not carried to the eagle's nest, where only two eaglets were seen surrounded by heaps of goat and sheep bones. It was not until two months later that a shepherd discovered the corpse of Marie Delex, frightfully mutilated, and lying upon a rock half a league from where she had been borne off.

A Tippah County, Mississippi, schoolteacher recorded a similar case in the fall of 1868. Noting that the eagles in the area had been very troublesome for some time, carrying off pigs, lambs, and other animals, she wrote that "no one thought that they would attempt to prey upon children." But "at recess, the little boys were out some distance from the [school]house, playing marbles, when their sport was interrupted by a large eagle sweeping down and picking up little Jemmie Kenney, a boy of eight years." When the teacher ran outside, "the eagle was so high" that rescue was impossible. The child died after the eagle eventually dropped him.

A tale that ended less tragically goes back to July 12, 1763, and the mountains of Germany. There a peasant couple left their three-year-old

THUNDERBIRD PHOTOGRAPH

In 1963 Jack Pearl wrote an article about a large bird-like creature in the men's action magazine *Saga*. Pearl insisted that in 1886 the *Tombstone Epitaph* had "published a photograph of a huge bird nailed to a wall. The newspaper said it had been shot by two prospectors and hauled into town by wagon. Lined up in front of the bird were six grown men with their arms outstretched, fingertip to fingertip. The creature measured about 36 feet from wingtip to wingtip." Unfortunately no one could produce the photo.

After receiving many letters and inquiries on the subject, the *Epitaph* conducted a complete search of its back issues but could find no such photograph. A far-reaching survey of other Arizona and California papers of the period brought the same empty results.

Janet Bord

(1945-)
and
Colin Bord

(1931-)

Janet Gregory met Colin Bord at a UFO study group in London in 1969. They were both interested in "earth mysteries: the study of prehistoric sites, folklore, [and] 'earth energies,'" according to Janet. Colin, a photographer, and Janet, an editor, were married in 1971. Together, the Bords developed the Fortean Picture Library for the *Fortean Times.* They have also written many books on a wide range of anomalies, from cryptozoology to UFOs to ancient British "earth mysteries" and folklore—like the history of the legendary Thunderbird.

About their approach to anomalies, Janet wrote: "We aim to be totally open-minded, though not gullible.... We know that mankind does not have all the answers and that there certainly are many mysteries which are little understood. Yet we try to maintain a commonsense approach to all mysteries, and we acknowledge that hoaxers are ever-present.... We have seen that some people become so immersed in the study of their chosen phenomenon that they cannot be objective.

"But above all, after 20 years of intense interest in mysteries and strange phenomena, we retain our youthful curiosity about all the anomalous happenings which are reported, we retain our sense of humor when an intriguing mystery is shown to have a prosaic [everyday] explanation, and we retain our intense interest in the human psyche, which after all is responsible for at least 50 percent of all mysteries."

In 1985 the Bords moved into an old stone house in a remote area of North Wales, where they continue to write about anomalies. One recent book, *Life Beyond Planet Earth?* (1991), examines the spectrum of evidence and folklore from astronomers' search for extraterrestrial life to the stories of flying-saucer contactees.

daughter lying asleep by a stream as they cut grass a short distance away. When they went to check on her, they were horrified to find her missing. A frantic search turned up nothing until a man passing by on the other side of the hill heard a child crying. As he went to investigate, he was startled at the sight of a huge eagle flying above him. And on the ground he found the little girl, her arm torn and bruised. When the child was returned to her parents, they and her rescuer figured that the bird had carried her well over 1,400 feet!

To twentieth-century zoologist C. H. Keeling, this eagle-kidnapping story is one of the most believable of the many on record. Still, he noted that *all* such stories seem to ignore the "simple and unalterable fact ... that no eagle on earth can carry off more than its own weight."

The Problem of Explanation

Most ornithologists (bird scientists) don't concern themselves with thunderbird reports, other than to reject them as foolishness. Illinois State University bird expert Angelo P. Capparella tried to explain why. "The lack of interest of most ornithologists in Thunderbirds is probably due to two factors," he wrote. "First, there is the lack of sightings from the legions of competent amateur birdwatchers.... [The] number of good birdwatchers scanning the skies of the U.S. and Canada is impressive. Every year, surprising observations of birds far from their normal range are documented, often photographically. How have Thunderbirds escaped their roving eyes?" A second reason, Capparella pointed out, is that such creatures have been reported in areas that lack the kind of constant food supply that the birds' huge appetites would require.

In addition, mistaken identifications have figured in thunderbird sightings. In some cases, witnesses have confused cranes, blue herons, and turkey buzzards with more extraordinary and mysterious birds.

Sources:

Bord, Janet, and Colin Bord, *Alien Animals,* Harrisburg, Pennsylvania: Stackpole Books, 1981.

Caras, Roger A., *Dangerous to Man: The Definitive Story of Wildlife's Reputed Dangers,* revised edition, South Hackensack, New Jersey: Stoeger Publishing Company, 1975.

Clark, Jerome, and Loren Coleman, *Creatures of the Outer Edge,* New York: Warner Books, 1978.

Coleman, Loren, *Curious Encounters: Phantom Trains, Spooky Spots, and Other Mysterious Wonders,* Boston: Faber and Faber, 1985.

Hall, Mark A., *Thunderbirds!: The Living Legend of Giant Birds,* Bloomington, Minnesota: Mark A. Hall Publications and Research, 1988.

Mount Ararat on Turkey's eastern border has been the site of many expeditions in the search for Noah's ark.

THE SEARCH FOR NOAH'S ARK

The Old Testament's Book of Genesis relates the story of Noah and his family, who, along with representatives of various animal species, escaped the Great Flood in a tremendous ark. After 40 days and 40 nights, the ark came to rest "upon the mountains of Ararat." Those who read the Bible as history (literalists and Christian fundamentalists, for example) place this event in the year 2345 B.C., though the Genesis account was recorded some 1,300 years later. Because of a lack of scientific evidence, however, most geologists and archaeologists doubt that any such worldwide flood ever took place.

In the view of many scholars, the story should be read as one of the many creation myths from all over the world about huge floods and chosen survivors. These narratives do not prove that a global flood occurred; rather, they are most likely tales of local floods, which, to victims, seemed to destroy all the world they knew.

To fundamentalists, however, such a view is not acceptable. So for a long time hopeful seekers have looked for the remains of Noah's ark on Mount Ararat. For if the Genesis story of Noah could be substantiated, this first book of the Bible's chronicle of other matters—particularly the Creation itself—could arguably also be trusted. There is in fact a Mount Ararat, or more specifically, two: Great Ararat (16,900 ft.) and Little Ararat (12,900 ft.). The peaks are connected by a rocky ridge between 7,000 and 8,000 feet high; they lie along Turkey's eastern border.

Because the name Ararat was not given to these mountains until roughly the eleventh century, many sources have placed the final resting place of the ark elsewhere. Most of the favored sites are in Turkey, but others include Greece, Armenia, and Iran.

Sightings and Searches

Despite shaky historical claims for an ark on Ararat, Jews and Christians have nonetheless continued to target this location. Around 1670 a Dutchman named Jan Struys, captured and enslaved by bandits in Armenia, reportedly met a hermit on Ararat. He treated the old man's illness, and in gratitude the hermit rewarded him with a "piece of hard wood of dark color" and a sparkling stone, both of which "he told me he had taken from under the Ark," according to Struys.

In the nineteenth century a number of searchers climbed the mountain but failed to find any sign of the ark. That is, until 1876, when James Bryce of England's Oxford University discovered a four-foot-long stick near the peak of Great Ararat. He declared it a piece of the ark.

On August 10, 1883, the *Chicago Tribune* published this report, which is clearly fascinating but almost certainly false:

> A paper at Constantinople announces the discovery of Noah's Ark. It appears that some Turkish commissioners appointed to investigate the avalanches on Mt. Ararat suddenly came on a gigantic structure of very dark wood, protruding from the glacier.... The Ark was in a good state of preservation.... They recognized it at once.

> There was an English-speaking man among them, who had presumably read his Bible, and he saw it was made of gopher wood, the ancient timber of the scripture, which, as everyone knows, grows only on the plains of the Euphrates. Effecting an entrance into the structure, which was painted brown, they found that ... the interior was divided into partitions 15 feet high.

> Into only three of these could they get, the others being full of ice, and how far the Ark extended into the glacier they could not tell. If, however, on being uncovered, it turns out to be 300 cubits long [the measurements cited in Genesis], it will go hard with disbelievers.

In 1892 Archdeacon John Joseph Nouri of the Chaldean Church reported that he had found the ark and even entered it. He also measured the structure, finding that it was indeed 300 cubits long.

During the following decades a number of ark expeditions were launched. Most ended in disappointment, though a few claimed sightings. A 1952 mission led by wealthy French industrialist Fernand Navarra produced samples of wood that, when first tested, dated back 5,000 years. But a later, more reliable test brought other findings: the wood was from A.D. 800 and was probably the remains of a monks' shrine built on the mountainside. When *Life* magazine published a photograph of a ship-shaped indentation in the mountain in 1960, an expe-

"It appears that some Turkish commissioners appointed to investigate the avalanches on Mt. Ararat suddenly came on a gigantic structure of very dark wood, protruding from the glacier."

dition raced there for an on-site look. The hollow turned out to be a natural formation created by a recent landslide.

Since then there have been other expeditions and claims, none especially notable. Most of the funding and personnel for these missions have come from fundamentalist groups. For them, the foundation of their religious beliefs is at stake: if the flood did not actually occur, and thus Noah and his ark did not exist, then other biblical pronouncements might also be questioned. Still, the findings of "arkeologists"—as ark investigators are called—have proven largely untrustworthy.

Critics have had no trouble finding flaws in arkeological thinking. Scientists Charles J. Cazeau and Stuart D. Scott, Jr., for example, remarked that "if the ark had come to rest near the summit of Ararat 5,000 years ago, it likely would have shifted by glacial movement to lower elevations long ago. To at least some extent, the ark would have broken up, the wood strewn about on the lower slopes of the mountain, easily accessible even to those who are not mountain climbers." Charles Fort, one of the first to compile accounts of strange and unexplained happenings, had this to say about the search: "I accept that anybody who is convinced that there are relics upon Mt. Ararat, has only to climb up Mt. Ararat, and he must find something that can be said to be part of Noah's Ark."

Sources:

Balsiger, Dave, and Charles Sellier, Jr., *In Search of Noah's Ark,* Los Angeles: Sunn Classic Books, 1976.

Cazeau, Charles J., and Stuart D. Scott, Jr., *Exploring the Unknown: Great Mysteries Reexamined,* New York: Plenum Press, 1979.

Fasold, David, *The Discovery of Noah's Ark,* London: Sidgwick and Jackson, 1990.

Other Strange Events

- SPONTANEOUS HUMAN COMBUSTION

- CATTLE MUTILATIONS

- CROP CIRCLES

- MAD GASSERS

Other Strange Events

SPONTANEOUS HUMAN COMBUSTION

Spontaneous human combustion (SHC) allegedly occurs when the heat inside a person's body becomes so great that he or she suddenly bursts into flames. The fire consumes both flesh and bones, leaving little more than ashes. Science has no explanation for the phenomenon.

One of the best-documented cases of SHC took place in 1951 in St. Petersburg, Florida, when a 67-year-old woman named Mary Reeser died under strange circumstances. Her body was discovered at 8 A.M. on July 2 when Reeser's landlady tried to enter her small apartment to deliver a message. The door handle was too hot to turn, so the landlady alerted two painters across the street, who broke down the door. Amid a great deal of smoke, they found Reeser's charred remains, reduced to ash except for one foot and what some accounts later claimed was a "shrunken skull" (the latter bit of information did not appear in the official report; a skull would ordinarily explode, not contract, in great heat). Ashes and a few coiled springs were all that remained of the overstuffed chair in which Mrs. Reeser had been sitting.

The incident attracted national attention and has been discussed often since in writings on SHC. But debate about such deaths goes back at least several centuries, and SHC was the subject of heated argument among eighteenth- and nineteenth-century medical practitioners. In a paper presented to the French Academy of Sciences in 1833, J. de Fontanelle reviewed a number of SHC cases and noted that victims tended to be old women fond of liquor. He also reported that fire damage did not extend to other flammable materials on or near the bodies. Reeser's nightgown, however, was incinerated, as was her chair.

> Amid a great deal of smoke, they found Mrs. Reeser's charred remains, reduced to ash except for one foot and what some accounts later claimed was a "shrunken skull."

The most famous
literary victim of
SHC was Mr. Krook
in Charles Dickens's
Bleak House;
illustration by Phiz.

In fact, for all the extraordinary claims surrounding Mary Reeser's death, some interpreted the evidence as pointing to a rather ordinary tragedy. When her son, a physician, saw her the evening before, Reeser told him that she had taken two sleeping pills. The official conclusion, that she fell asleep while smoking and burned up along with her flammable nightgown, is not unreasonable. Apparently, her own body fat— Reeser was overweight—further fed the flames.

One popular pre-twentieth-century theory about SHC was that those prone to alcohol abuse were particularly at risk—because the liquor in their systems made them more burnable. (The most famous literary victim of SHC was Mr. Krook in nineteenth-century British novelist Charles Dickens's *Bleak House.* Krook was intoxicated at the time of his extraordinary combustion.) There is no scientific basis for this connection, though there is a relationship between drunkenness and fire deaths. "Drunken persons," Joe Nickell and John F. Fischer wrote, are "more careless with fire and less able to properly respond to an accident."

In several articles questioning SHC claims, Nickell and Fischer noted that suspicious items were often found at the sites: "a broken oil lamp on the floor, a victim's pipe, a candlestick lying near the remains. But in addition there was often a large quantity of combustible material under the body to aid in its destruction: bedding, for example, or a chair's stuffing—even wooden flooring possibly impregnated with oils or waxes. Interestingly enough, there was evidence that melted human fat had increased the destruction in a number of instances."

"Protestations of the debunkers notwithstanding, the case *for* SHC is actually very good," wrote Larry E. Arnold, director of ParaScience International in Harrisburg, Pennsylvania. Arnold has investigated this hotly debated subject for 20 years.

Fires kill thousands of people each year, Arnold pointed out, yet rarely do firemen find victims reduced to ashes amid largely unsinged surroundings. "Why not?," he asked. "If dropping a cigarette or pipe on one's lap could transform a hapless human to ash within minutes, this would happen thousands of times annually. It doesn't. Something *else*—some thing(s) far less obvious—must cause a person to become a human fireball."

Firemen have told Arnold that classic SHC is unforgettable because its characteristics are so unusual. First: incineration of the victim can be more thorough than in a crematorium. Second: accelerants (like gasoline or kerosene, even alcohol) needed to fuel such an intense blaze are absent. Third: fire damage is incredibly localized—often nearby newspapers are unscorched, plastics unmelted. Fourth: rather than the terrible odor of burned flesh, a sweet "perfume" smell is sometimes detected.

> In a crematorium, a skeleton becomes wholly burned to dust only in heat above 3,000 degrees Fahrenheit for 12 hours--much hotter and longer than the worst house fire.

Dr. John Bentley's baffling burning in 1966 featured each of these mysterious traits. Overnight, the 92-year-old physician defied common sense and science when he burned through a highly flammable linoleum floor, leaving behind but half a leg, a mound of ashes, and a sweet smell. Yet paint on his bathtub only inches away did not blister, and rubber tips on his aluminum walker (with a melting point of only 1,200 degrees Fahrenheit) did not melt! No wonder a Pennsylvania State Police officer murmured that Dr. Bentley had died by "spontaneous human combustion."

The remains of Dr. John Bentley—a victim of SHC?

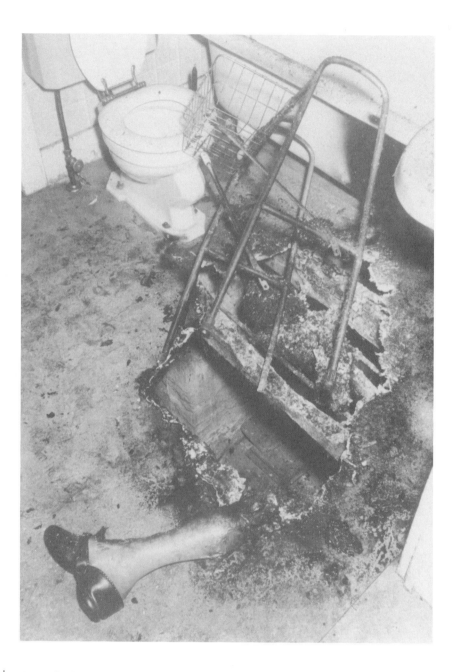

Partial SHC

More astounding, SHC is not always fatal! In 1974 Jack Angel, a traveling salesman, parked his motor home at a Georgia motel and went to sleep. He awoke to a real-life nightmare: his right hand and forearm were burned black. Amazingly, nothing around him was singed, not his pajamas, not even the sheets on which he slept! Puzzled physicians diagnosed his burns as "internal in origin"—that is, he burned from the inside out. They likened his injuries to a high-energy electricity burn, but no one could detect the source of the electricity, unless it came, suggested Arnold, from the electrical potential that Angel—and everyone—has naturally inside his or her own body.

SHC has even been witnessed. In 1980 Peter Jones of California *twice* watched smoke wisp out of his body—once from his forearm while driving his car and, later, from his feet as he was about to slip on his slippers one morning. His wife saw the second incident herself. "What was that? Were you smoking?" she cried out. He was, but not with cigarettes! Jones, like Angel, lived to tell an incredible story of surviving his own inflaming.

"That some strange fiery fate has befallen selected individuals throughout history is undeniable, *if* the evidence is examined," stated Arnold. "Inquiring minds will one day find the explanation to this phenomenon."

Sources:

Arnold, Larry E., "The Flaming Fate of Dr. Bentley," *Fate* 30,4, April 1977, pp. 66-72.
Arnold, Larry E., "The Man Who Survived Spontaneous Combustion," *Fate* 35,9, September 1982, pp. 60-65.
Harrison, Michael, *Fire from Heaven: A Study of Spontaneous Combustion in Human Beings,* London: Sidgwick and Jackson, 1976.
Nickell, Joe, and John H. Fischer, *Secrets of the Supernatural: Investigating the World's Occult Mysteries,* Buffalo, New York: Prometheus Books, 1988.

CATTLE MUTILATIONS

In the fall of 1973, farmers in Minnesota and Kansas reported that their cattle were dying mysteriously. It appeared that unknown forces had killed the animals, but without knives or bullets. Worse still, various body parts—usually eyes, ears, lips, sex organs, rectums, and tails—had been removed with surgical skill. Farmers also frequently claimed that the animals' blood had been drained. And strangest of all was the fact that the killers did all of these things without leaving footprints or any other signs of their presence behind.

Farmer looks at cow—one of many alleged cattle mutilations across Middle America and Canada.

Law officers were mystified. According to Deputy Gary Dir of Ottawa County, Kansas, "The large majority of these mutilations occurred near occupied houses. In no instances were the animals found less than a quarter-mile from the roadside and none ... more than a quarter-mile from an all-weather, well-traveled road." One carcass in Cloud County was discovered in a mud hole. Even so, there were no footprints!

In December a dozen Kansas sheriffs met to discuss the problem. Though they had little to go on, most agreed that members of a satanic cult were probably responsible for the killings. In southwestern Minnesota, however, officials disagreed. Lincoln County Sheriff Albert Thompson, who had investigated several of the mutilations, was certain that the animals had died of ordinary cattle diseases and that the so-called cuts were left by small animals that had chewed on the soft parts of the bodies. Regardless, many country people remained convinced that a group of Satan worshippers rumored to operate in the area, had killed the animals in bizarre sacrificial ceremonies.

When Kansas authorities brought the carcasses to the Kansas State University Veterinary Laboratory, pathologists found that the cause of death was blackleg, a bacterial disease often fatal to cattle. State brand commissioner Doyle Heft assured all that nothing out of the ordinary was afoot. Still, this did not put an end to wild theories about such "cattle mutilations."

Whether real or imagined, cases of cattle mutilation were being reported in other states in the Midwest and West and even into Canada's western provinces by 1974. And by the end of the decade, newspapers claimed that several thousand unexplained cattle deaths had taken place, with the killers' identities still unknown. Fear and imagination began to run rampant. Four predominant theories explaining the mutilations emerged. The first blamed cultists; a second alleged a government conspiracy in which agents were conducting secret chemical/biological-warfare experiments; a third pointed to UFOs and space beings; and the fourth blamed mass panic for what were, in fact, commonplace deaths.

The Satanist Theory

Police agencies in Idaho, Montana, and Alberta, Canada, did uncover a few cases in which satanist groups could be linked to cattle mutilations. Laboratory studies indicated that a small number of animals had been killed after being drugged. In Idaho a police informant infiltrated a group that claimed to mutilate cattle, though he did not personally witness such an act. Some reliable sightings of black-hooded figures were recorded, but without proof of their connection to animal deaths. And police officers, farmers, and ranchers sometimes stumbled on what they believed were signs of cult activity, such as stone altars and bodies of small animals.

Still, in 1975 Donald Flickinger, a Minneapolis-based agent of the U.S. Treasury Department's Alcohol, Tobacco and Firearms division, was assigned to investigate reports of a nationwide satanic network involved in animal and human mutilation. He found no supporting evidence.

Government Conspiracies

Theories about secret government involvement in cattle mutilations were not surprising during the 1970s, the era of the Vietnam War and the Watergate break-in, events that shattered the faith many Americans had in their government. Nonetheless, such notions were rarely based on proof. The only physical evidence to support this idea came to light in Lincoln County, Colorado, in 1975. A rancher found a blue

A rancher found a blue government bag near his mailbox. Inside he discovered plastic gloves, a bloody scalpel, a cow's ear, and part of a tongue.

government bag near his mailbox. Inside he discovered plastic gloves, a bloody scalpel, a cow's ear, and part of a tongue. The Colorado Bureau of Investigation could find no fingerprints on the items, nor could local law officers connect the animal parts with any cattle-mutilation reports known to them.

Reports of "mystery helicopters" also inspired government conspiracy hypotheses regarding cattle mutilations. Examining a number of sightings, "mutologists" Tom Adams and Gary Massey remarked that such aircraft "are almost entirely without identifying markings, or markings may appear to have been painted over or covered with something. The craft are frequently reported flying at abnormal, unsafe, or illegal altitudes. The mystery choppers may shy away if witnesses or law officers try to approach. On the other hand, there are several accounts of aggressive behavior on the part of the helicopter occupants, with witnesses chased, 'buzzed,' hovered-over or even fired upon." No direct link between these reports and cattle mutilations has ever been made, however.

Evil Aliens

To a number of mutologists, farmers, ranchers, and country police officers, the extraordinary way in which the cattle were killed—with supposedly precise incisions and with no footprints left behind—suggested mutilators from an extraordinary place: outer space. *Strange Harvest,* a documentary written and produced by Denver filmmaker Linda Moulton Howe in 1980, explored this idea. It attracted a great deal of attention and shaped popular beliefs about UFO-connected cattle mutilations.

Belief in cattle-killing space beings spread quickly, even without credible evidence. Of course, few UFO reports suggested a definitive connection to cattle deaths. One that did was related under hypnosis: a woman told University of Wyoming psychologist and ufologist R. Leo Sprinkle that she had seen a cow drawn up into a UFO "in a pale, yellow beam of light." She and her daughter were also taken inside the vessel, where they saw aliens cutting up the animal. Shortly afterwards, Sprinkle hypnotized a second woman who told a somewhat similar story.

These accounts, among others equally improbable, inspired a complex conspiracy theory that drew quite a following in the early 1990s. It was based on the belief that evil UFO beings had entered into an agreement with America's "secret government," which permitted the aliens to abduct and mutilate cattle and even to abduct human beings

Other Strange Events

in exchange for extraterrestrial technology. In some versions, the government even allowed the mutilation of people. These wild stories, completely unsupported by evidence, have been spread through books, lectures, and videos by conspiracy theorists Milton William Cooper, William English, and others. All claim to have learned these terrible truths from unnamed government informants and secret documents.

Ordinary Causes

In 1979 the First Judicial District of New Mexico received a federal grant to investigate mutilations in that state. Former FBI agent Kenneth Rommel headed the study, and at its conclusion in April 1980, the report found no evidence of cattle mutilations. Rommel had worked on 24 cases in New Mexico and had kept in close contact with law officers while addressing reports in other states. He felt that all of the mutilations he investigated were ordinary, "what one would expect to find from normal predation [large animals hunting for food], scavenger activity [smaller creatures feeding on remains], and decomposition [decay] of a dead animal."

Rommel blamed faulty investigation, guesswork, unchecked imagination, and blatant stupidity for creating a mystery where there was none. His conclusions matched those of investigators in many other states but received more attention because he published them into a detailed official report. Social scientists who studied the cattle-mutilation panic viewed it as a case of mass hysteria born of exotic theories and unproven statements that were spread—without question—by the press.

In 1984 New York writers Daniel Kagan and Ian Summers wrote *Mute Evidence,* a careful, well-researched book on the subject. Examining the origins and development of the panic, Kagan and Summers found that a small group of "mutology" buffs, most of them also UFO enthusiasts, were to blame. According to the authors, none of these people:

> had access to any experts in veterinary medicine, livestock, or any other fields that bore on the cattle mutilation question, and it was obvious there was not one seriously qualified investigator in their underground. They were all amateurs, all poorly trained to deal with the subject, and all seemingly uniquely ignorant of research procedures and methods of constructing proven cases.... They had nothing going for them, yet they controlled the opinions of literally hundreds of thousands, perhaps millions of people, regarding cattle mutilations.

By the early 1980s press accounts claimed that as many as 10,000 mutilations had taken place, but Kagan and Summers knew better; they had checked official cattle mortality reports and found that cattle had died at a statistically average rate throughout the most intense years of the mutilation scare. The 10,000 figure was revealed as the invention of a mutologist who admitted that he had pulled the number out of thin air!

In 1991 an Arkansas newspaper reported on a "mutilated" heifer calf that a veterinarian determined had died of blackleg, the corpse of which had been disturbed by buzzards. Also included in the article was a quote from two "UFO investigators" who said that a whopping 700,000 mutilations had occurred and that alien beings had used "lasers" to do the cutting, evidence that even in light of the facts, tall tales of cattle-mutilation endure.

Sources:

Bayles, Fred, "Scoffers, Believers Abound in Mutilated-Cattle Mystery," *Washington Post,* January 1, 1986.

Howe, Linda Moulton, *An Alien Harvest: Further Evidence Linking Animal Mutilations and Human Abductions to Alien Life Forms,* Littleton, Colorado: Linda Moulton Howe Productions, 1989.

Kagan, Daniel, and Ian Summers, *Mute Evidence,* New York: Bantam Books, 1984.

CROP CIRCLES

Crop circles first attracted public attention in the early 1980s, when circular patterns were found in crops of growing grain in the countryside of southern England. Since then they have increased in both number and complexity, and the term now refers to a variety of patterns: from simple single circles to quintuplets (a central circle ringed by four smaller ones), to dumbbell shapes and combinations of these involving lines, bars, ladder-like rungs, and more. These intricate patterns are called "pictograms" because they resemble primitive rock paintings.

"Cereology," the study of crop circles, arose after the *Wiltshire Times* (August 15, 1980) published an article and photographs featuring circles found flattened in a field of oats in Bratton, Wiltshire, England. Each was about 60 feet across and swirled flat in a clockwise direction. The Bratton circles stirred the interest of both meteorologist George

Terence Meaden, of the Tornado and Storm Research Organization (TORO), and ufologist Ian Mrzyglod. Mrzyglod made two important discoveries: the circles showed no clear signs of radiation, and they were not really circles at all; the formations were slightly elliptical, an unexpected finding that seemed to argue against a hoax.

A year later, on August 19, 1981, three more circles were found in neighboring Hampshire County, alongside a main highway. Where the Bratton circles appeared random, these at Cheesefoot Head looked as if they had been laid out along a straight line. On either side of the main circle—again 60 feet across—were smaller circles of 25 feet. All were swirled clockwise.

The crop circles seemed to follow a pattern; they appeared mostly during the spring and summer growing season in the rolling grain fields west and southwest of London. This "enchanted" landscape was already home to other archaeological mysteries, including the monoliths of Stonehenge and Avebury, the pyramid-like peak of Silbury Hill, and carvings cut in the chalk hills.

While there is no public database recording the number and types of crop circles that appear yearly, the best available information suggests that during the years 1980 to 1987, between 100 and 120 circles were formed. During that time they also displayed several "mutations." Some circles were swirled counterclockwise; sometimes rings appeared around them. And crop circles varied in size, becoming a great deal larger or smaller.

In 1988 at least 112 circles were recorded, matching the combined total of the eight previous years. In 1989 the number almost tripled, to 305; it tripled again in the summer of 1990 to about 1,000. In 1991 there were 200 to 300 recorded circles, many of them the more complex pictogram type.

By the early 1990s, what had begun with a few simple circles a decade before had mushroomed into a mystery of worldwide scope, with well over 2,000 circles recorded around the globe. Similar circles—though rarely as numerous or complex as those found in England—were now noted in the Soviet Union, the United States, Canada, Australia, Japan, and other countries.

Generally, a crop circle occurs overnight, probably in the hour or two before dawn. It apparently happens in a matter of seconds, usually 60 or less.

Characteristics

Generally, a crop circle occurs overnight, probably in the hour or two before dawn. It apparently happens in a matter of seconds, usually

A typical "simple" crop circle near Milk Hill, Wiltshire, England.

60 or less. The line between the affected and non-affected crop is almost always abrupt and dramatic. The flattened area is laid down in a spiral manner from the center outward. In addition, the crop is frequently flattened in layers lying in opposite or differing directions. Once in a while the plant stalks even appear to be braided or intertwined with one another. If a crop circle forms early in the growing season, the affected plants continue to grow and will "bounce back" to nearly normal height. Flattened later in the year, however, the stalks remain on the ground.

The stalks involved are laid down without breaking and show no signs of damage. This is true of even delicate plants, like rape, the source of canola oil. The plant stalks seem to go limp, almost as if they had been steamed or made more elastic. An encircling ring may

Meteorologist George Terence Meaden (kneeling) shows how the floors of some crop circles are laid down in layers. This particular formation appeared near Bishops Cannings, Wiltshire, in June 1990.

run counterclockwise to the central circle or vice versa, or both may be laid in the same direction. The smallest crop circle on record measured just eight inches across. And at Alton Barnes in 1990, a pictogram of complex crop circles stretched for nearly an eighth of a mile!

Witnessing the formation of a crop circle is rare. There are roughly three dozen such reports, most collected by Dr. Meaden and his associates at the Circles Effect Research Group (CERES), which is part of TORO. Typical is the account given by Gary and Vivienne Tomlinson published in the *Mail on Sunday* on August 25, 1991, a year after the event took place. The Tomlinsons were walking alongside a field of grain near the village of Hambledon when the plants to their right suddenly started rustling.

"There was a mist hovering above, and we heard a high-pitched sound," said Mrs. Tomlinson. "Then we felt a wind pushing us from the side and above. It was forcing down on our heads so that we could hardly stay upright; yet my husband's hair was standing on end. It was incredible. Then the whirling air seemed to branch into two and zig-zagged off into the distance. We could still see it like a light mist or fog, shimmering as it moved."

"Then we felt a wind pushing us from the side and above. It was forcing down on our heads so that we could hardly stay upright; yet my husband's hair was standing on end."

Crop Circles

Meaden was particularly interested in the Tomlinson account because it seemed to support his own ideas about crop circles. He felt that the simple ones, at least, were formed by the breakdown of a standing, electrically charged whirlwind or plasma-vortex. Unlike regular whirlwinds such as dust devils and waterspouts, which suck in surrounding air, dust, or water at the base of a tunnel of rising air, Meaden's plasma-vortex falls apart, or collapses, in a descending burst of violent wind. It is this collapsing wind-form, surrounded at times by a ring of electrically charged air, that quickly cuts out crop circles. The meteorologist also felt strongly that the low-lying hills of southern England provided the perfect physical conditions for formation of these unusual whirlwinds.

Have crop circles occurred throughout history? This is an important consideration in any attempt to explain them. If crop circles can be established as a regular feature of the natural environment, then a weather-based theory like Meaden's makes sense. On the other hand, the theory is weakened if crop circles are shown to be a uniquely modern occurrence. How crop circles might have been reported in the past, then, is an issue worth investigating.

Fairy Rings

Unexpected circular shapes in crops or grasses have long been associated with so-called fairy rings. The typical fairy ring is caused by a mushroom fungus, *Marasmius oreades,* and has two stages. In the first, the fungus decomposes, or breaks down, living material, which spurs the growth of plants in the affected area; in the second, these plants become so abundant that all growth is choked out, resulting in the bare appearance of a ring worn down by the patter of many little feet—ostensibly those belonging to fairies.

But fairy rings are not as clear-cut and complex as crop circles. And they demonstrate either the presence or absence of growing plants, whereas the affected stalks of a true crop circle have simply "fallen over" in place. Still, it is useful in understanding crop circles to examine historical accounts of fairy and other rings from a modern viewpoint, weeding out old-time beliefs and superstitions.

John Aubrey's *Natural History of Wiltshire* relates an intriguing seventeenth-century account of fairy mischief that reportedly befell a clergyman named Mr. Hart employed by the Latin School at Yatton Keynal. He told a student that one evening he was particularly annoyed with the elves and fairies that frequently came down from the hills at dusk to sing and dance. As Hart approached one of "the green circles made by those spirits on the grass [called 'fairy dances' by the local people], he all at once saw innumerable small people dancing round and round, singing, and making all manner of small odd noises." Hart, "very greatly amazed," found that he could not "run away from them," feeling he was "kept there in a kind of enchantment." As soon as the fairies noticed him, they "[surrounded] him on all sides, and ... he fell down scarcely knowing what he did; and thereupon these little creatures pinched him all over, and made a sort of quick humming noise all the time; but at length they left him, and when the sun rose he found himself exactly in the midst of one of these fairy dances."

Compare this account with the Tomlinsons', reported three and a half centuries later, in the same general location. The fairy incident occurred in August, as did the Tomlinsons'. The humming sound Hart heard (minus the "pygmies" of course!) could be likened to the high-pitched sound noted by the couple. Could Hart's feeling of being "pinched ... all over" be the prickling sensation caused by static electricity or some other electromagnetic effect? Could his inability to run away and later, to remain standing, be the result of a strong wind?

Saucer Nests

Crop circles also recall the saucer nests mentioned in UFO writings: circular indentations that one could imagine to be left by hovering or grounded spacecraft. Two of the most notable saucer-nest cases took place at Tully, Queensland, Australia, in January 1966, and near Langenburg, Saskatchewan, Canada, on September 1, 1974. Both involved daytime UFO sightings. The Langenburg event is worth noting for several reasons: not only did it occur before the crop-circle craze, but it involved a reliable witness, photographs, and investigations by both the Royal Canadian Mounted Police and Ted Phillips, a Center for UFO Studies researcher then specializing in physical-trace cases.

At about 11 A.M. that September Sunday, 36-year-old Edwin Fuhr was cutting his family's rape crop with a mechanical harvester. The day was overcast and cool, with a falling mist and light showers in the area. Nearing the end of the field, Fuhr noticed what appeared to be a metallic dome some 50 feet away. He left his machine idling and approached the object on foot, coming to within 15 feet. It now looked like an upside-down stainless steel bowl, 11 feet across and about five feet high, and was spinning rapidly. It appeared to be hovering 12 to 18 inches above the ground.

Fuhr returned to his harvester. Now he could see four similar objects nesting nearby, arranged in a semicircle, all rapidly rotating. Because his machine was still running, he could not tell whether the spinning domes were emitting a sound or not. Then all five objects suddenly rose in the air to a height of about 200 feet, where they hovered in stair-step formation, no longer spinning. Eventually, each gave off a puff of gray vapor or smoke and vanished into the low clouds. Fuhr told Phillips that their departure created a "pressure that flattened the rape that was standing, and I thought, 'Oh, hell, here goes my crop,' and there was just a downward wind, no twirling wind. I had to hold onto my hat."

Though confused, Fuhr felt the episode lasted about "15 to 20 minutes." He also reported moments of paralysis during the experience—he simply could not move.

After the objects left, Fuhr found five circular areas spiraled flat in a clockwise direction. "I checked for burns," he said, "but I couldn't find any. *The grass wasn't broken off, it was flat, pressed down.* It didn't seem different from the other grass."

Other strange effects frequently mentioned in UFO close-encounter writings were also noted in the area. At about the same time as Fuhr's experience, the cattle in a neighboring field broke through their fencing in four places. The night before, Fuhr's own farm dogs had barked wildly. On Monday night they acted up again, and the following morning a sixth circle was found in the field. On the night of September 14, his dogs did the same, and a seventh circle was found on the Fuhr farm the next morning.

Still, the connection that fairy rings and saucer nests have to today's crop circles is shaky at best. But there have been reports of nocturnal (night) lights appearing at locations where crop circle formations have later been discovered. And at least two daytime videos of small lights sailing near or diving into existing circles have been recorded. These unknown lights further complicate the mystery.

Physical Effects

It has been reported that people standing close to crop circles have felt a number of physical effects. While such personal sensations are difficult to analyze, the Center for Crop Circle Studies has put together a record of these accounts. The most commonly cited aftereffect of visiting a crop circle is nausea, followed by headache and extreme fatigue. These symptoms last only briefly. On the other hand, some circle visitors tell of pleasant feelings.

The failure of electronic equipment in the presence of crop circles has been detected a number of times. This is especially true of still and video cameras, but also includes audio tape recorders. Once, circle investigators Colin Andrews and Pat Delgado recorded high-pitched warbling sounds while in the company of a British Broadcasting Company video crew.

Close scientific study of plant stalks from crop circles has opened up another area of investigation. American bio-physicist W. C. Levengood found that the nodes or knuckles of affected stalks appeared larger—slightly swollen—compared with normal samples. Examination of the stalks' microscopic cell-wall holes, through which nutrients pass, revealed stretched or larger pits—a sign of rapid microwave heating. In addition, seeds from affected stalks seemed to sprout more quickly than regular seeds. More testing in these areas is needed.

Hoaxes

Some crop circles are believed to be hoaxes. The question remains: how many and which ones? While some cereologists accept pictograms and other complex patterns and formations as genuine, those who believe in a meteorological explanation for the circles feel that anything beyond the simplest shapes are suspicious. Researchers have tried to reach an agreement on guidelines and methods for determining which crop circles are genuine. (Energy "dowsing"—using a divining rod—has been a frequently employed method, for example, but it is highly questionable.)

In most cases, "authentification" still depends on a ground-level, visual inspection of the formation and the experience of the investigator. Sometimes flying over a crop formation is helpful because hoaxes often appear crude or ragged from above.

Probably the biggest crop-circle hoaxers to have come forward are Doug Bower and David Chorley, two elderly Englishmen who claimed to have created some 250 complex formations. On September 9, 1991, the British tabloid *Today* published their detailed confessions. According to the pair, they began their deception in the summer of 1978 with a simple circle near Cheesefoot Head, Wiltshire, that was easily seen

How many—and which—crop circles are hoaxes?

from the road. Bower, who had lived in Australia from 1958 to 1966, said he got the idea from the saucer nests that had appeared during that time in Queensland. "We had a good giggle about the first one," Chorley recalled. "It was nice being out on a summer night, so we decided to do some more. But for three rotten years [the papers] never noticed what we were doing."

Bower and Chorley said that once the press and public did take note, they improved their methods and created more complex formations. Frequently they would include their initials—in the form of a double-D—in their handiwork. The two claimed that they finally came forward because others (like Andrews and Delgado, coauthors of two best-selling books on the subject and founders of Circles Phenomenon Research) were profiting from their secret efforts. With *Today*'s help, they created a complex crop formation and invited Delgado to inspect it. Hoaxers Bower and Chorley at last became famous when the investigator declared their formation genuine.

Cereologists were embarrassed by Delgado's mistake. But they challenged Bower and Chorley again. The two made a second daytime circle before the media using the simple tools—string, rope, four-foot-long wooden planks, and a crude sighting device—that they claimed to have used in their early creations. The result was ragged and poorly constructed. (Perhaps more importantly, Bower and Chorley have yet to demonstrate their ability to create a complex crop-circle formation at night, when most appear.) Other groups, including the local Wessex Skeptics, have also created crop circles that have fooled experts.

Yet questions remain about human involvement. Some formations—such as the immense pictogram that appeared at Alton Barnes—are constructed on an enormous scale. Assuming that this and similar formations are hoaxes, why has no huge crop circle ever been discovered interrupted or abandoned—for whatever reason—halfway through completion?

In at least one well-documented case in the summer of 1991, Meaden and a team of visiting Japanese scientists were watching a field with electronic equipment that included radar (sound waves), magnetometers (which measure magnetic force), night-vision video cameras, and motion sensors. Blanketed by mist, a small dumbbell formation appeared; yet none of the sensing equipment noted intruders! In the years since crop circles first appeared, farmers and landowners in the affected areas have watched their property more closely than ever. But the number of hoaxers caught has remained quite small.

Competing Theories

UFOs, secret military experiments (that produce microwave or laser radiation), and psychokinesis (the movement of objects with the mind alone) have also been named as causes of crop circles. Doubters, of course, blame human activity—in other words, all crop-circle formations are hoaxes.

DOWSING AND CROP CIRCLES

Meaden's concept of a plasma-vortex, a hypothetical collapsing wind-form that could cut crop circles in a descending burst of violent wind, remains the only "mainstream" weather-based theory to offer an explanation for the formation of crop circles. Yet many cereologists feel that it simply does not account for all of the reported "behaviors" of the phenomenon. Some think that an unknown natural force or intelligence within the earth may be behind them. Richard Andrews, a dowser with the Center for Crop Circle Studies, has claimed that crop-circle patterns, or their energy fields, can be *dowsed* a year or more before the actual circle appears.

Dowsing is a folk method of finding underground water or minerals with a divining rod. The divining rod is usually a forked twig; the "diviner" holds the forked ends close to his body, and the stem supposedly points downward when he or she walks over the hidden water or desired mineral. Many believe that luck is behind most dowsing successes.

Whatever their origins, there is no doubt that English crop circles have captured the imagination of a curious public. For "unlike ghosts, poltergeists, or even UFOs," explained author Hilary Evans, "the circles are absolutely there for anyone to examine at will." Organizations, newsletters, books, videos, tours, and more have sprung up around them. With so much attention focused on crop circles from both inside and outside the scientific community, can a breakthrough be very far behind?

Sources:

Delgado, Pat, and Colin Andrews, *Circular Evidence,* London: Bloomsbury, 1989.
Delgado, Pat, and Colin Andrews, *Crop Circles: The Latest Evidence,* London: Bloomsbury, 1990.
Meaden, George Terence, "Circles in the Corn," *New Scientist,* June 23, 1990, pp. 47-49.
Meaden, George Terence, ed., *Circles from the Sky,* London: Souvenir Press, 1991.
Noyes, Ralph, ed., *The Crop Circle Enigma,* Bath, England: Gateway Books, 1990.
Randles, Jenny, and Paul Fuller, *Crop Circles: A Mystery Solved,* London: Robert Hale, 1990.

MAD GASSERS

In late summer of 1944, a mysterious assailant terrorized the small east-central Illinois town of Mattoon (population: 15,827). For two weeks townspeople reported frightening episodes of gassings in their homes. Their attacker would eventually be declared imaginary by local law officials, who considered the case one of mass hysteria. Historians and psychologists who later reviewed the case concurred with this conclusion. Still, unanswered questions about the Mattoon episode remain.

Mattoon's phantom-like menace first made his or her presence known on August 31 when a resident woke up feeling ill. He managed to get to the bathroom before throwing up. Returning to the bedroom, he asked his wife if she had left the gas on. She said she hadn't. But when trying to get up to check, she found she couldn't move. Elsewhere in town a young mother who heard her daughter coughing in another room also tried to leave her bed, but she too experienced this paralysis.

At 11 P.M. on September 1, a "sickening sweet odor in the bedroom" awoke another young mother, Mrs. Bert Kearney, from sleep. As the odor quickly grew stronger, she explained, "I began to feel a paral-

An artist's imaginative drawing of Mattoon's "mad gasser" depicts him as a not-quite-human, possibly extraterrestrial, being.

ysis of my legs and lower body. I got frightened and screamed." An hour and a half later, when her husband came home from work, he saw a strange man standing at the bedroom window. Kearney described him as "tall, dressed in dark clothing and wearing a tight-fitting cap." Kearney chased the prowler, but the man escaped.

These events took place before anyone had heard of a "mad gasser"; thus mass hysteria could not have been a factor. But the *Mat-*

toon Journal-Gazette soon picked up the story, covering it in a sensational way that frightened more than it informed. The newspaper also hinted that more incidents would follow; when further attacks did occur, they were illustrated in the same dramatic (and one could say irresponsible) manner.

Several other residents complained to police that the sudden flow of a "sickly sweet odor" into their homes had paralyzed them for as long as 90 minutes. No one else had seen the gasser, but late on the evening of September 5 as one couple was returning home, the wife noticed a white cloth by the front door. When she picked it up, she happened to sniff it. "I had sensations similar to coming in contact with an electric current," she recalled. "The feeling raced down my body to my feet and then seemed to settle in my knees. It was a feeling of paralysis." Soon her lips and face were burning and swelling, her mouth was bleeding, and she was vomiting.

These dramatic symptoms had passed by the time police arrived, but once at the scene, officers did discover what seemed to be the first physical evidence in the case: an empty lipstick tube and a skeleton key (a kind of filed-off master key that can open numerous locks) near where the noxious white cloth had lain. Even as officers were interviewing the couple, however, a woman elsewhere in town was hearing a prowler outside her bedroom window. Before she could sit up, a gas seeped into the room, and she was unable to move for several minutes.

Near midnight a woman called police to report that a man had tried to force his way through her door. Her screams frightened him off. Press accounts of the incident described the man as the "mad gasser." Despite the uncertainty of the identification, this was just the kind of story to fuel a growing panic. Two nights later, a woman and her 11-year-old daughter said they heard someone attempting to break open a window. They tried to get outside, but mysterious fumes overcame the mother and made her sick.

In a September 8 article on the gassings, the Decatur *Herald* noted: "Victims report that the first symptom is an electric shock which passes completely through the body. Later nausea develops, followed by partial paralysis. They also suffer burned mouths and throats and their faces become swollen."

No End in Sight

As the days passed and the attacks continued, Mattoon residents were outraged that local police had not been able to catch the gasser.

> Several other residents complained to police that the sudden flow of a "sickly sweet odor" into their homes had paralyzed them for as long as 90 minutes.

A protest rally was planned. Armed citizens prowled the streets at night, ignoring the police commissioner's plea that they behave rationally. He admitted that a "gas maniac exists," but added that "many of the attacks are nothing more than hysteria. Fear of the gas man is entirely out of proportion to the menace of the relatively harmless gas he is spraying." Rumor had it that the gasser was a lunatic or an eccentric inventor.

The scare peaked on September 10, with two attacks striking five people. But by the next morning police were skeptical, pointing to a lack of solid evidence; they decided that all further "victims" would undergo physical and psychological testing. A chemical study of the cloth found five days earlier revealed little. The next evening, when the police received more calls from people reporting attacks, they dismissed them as false alarms (though in one case a physician who went to a victim's house smelled the gas himself).

At a press conference on the morning of September 12, the police chief told reporters: "Local police, in cooperation with state officers, have checked and rechecked all reported cases, and we find absolutely no evidence to support stories that have been told. Hysteria must be blamed for such seemingly accurate accounts of supposed victims." Beyond that, he theorized, the odor of carbon tetrachloride from a nearby chemical plant may have been carried on the wind. He did not explain why this had never been a problem for Mattoon residents in the past.

Even in the face of this official denial, the gasser made one last house call. On the evening of the 13th, a witness saw a "woman dressed in man's clothing" spray gas through a window into Bertha Burch's bedroom. The next morning Mrs. Burch and her adult son found footprints of high-heeled shoes under the window.

The Botetourt Gasser

In 1945, writing in the *Journal of Abnormal and Social Psychology,* Donald M. Johnson reviewed the Mattoon scare and concluded that the local newspaper's alarming coverage was responsible for the phenomenon from start to finish. Johnson's study would influence future investigations of mass panics.

But unknown to Johnson and most other examiners of the Mattoon episode, a strikingly similar series of events took place in Botetourt County, Virginia, in December 1933 and January 1934. The scare earned only local coverage, and it is unlikely that Mattoon residents were aware of it.

The first recorded Botetourt attack occurred at a farmhouse near Haymakertown late on the evening of December 22 when three separate gassings sickened eight members of a family and a visitor. Some of the victims thought they saw a man fleeing in the darkness. The gas caused nausea, headaches, facial swelling, and tightening of the mouth and throat muscles. One victim, a 19-year-old woman, suffered convulsions for weeks afterwards. A police officer who investigated found only one clue: the print of a woman's heel under the window where the gas was believed to have entered.

Over the next two weeks other people reported similar night attacks. In one case witnesses saw a 1933 Chevrolet with a man and a woman inside driving back and forth in front of a house around the time its occupants experienced a gassing. In another instance a young mother attending to her baby said she heard a rattling window shade and mumbling voices outside. Suddenly the room filled with gas, and her body felt numb. While on his way to call police after a gassing at his farm, F. B. Duval saw a man run toward a car parked on a country road and quickly drive away. Duval and an officer examined the site soon afterwards and found prints of a woman's shoes.

Amid growing panic, residents of the county armed themselves and prowled back roads in search of suspicious strangers. One even fired at a fleeing figure, who nonetheless eluded the shot. Another time, moments after a gas attack, one of its victims dashed outside in time to glimpse four men running away. By the time the witness returned with a gun, he could no longer see them, but he could still hear their voices. Of course, there were those who doubted that such attackers existed. But physicians who treated victims were certain that the gassings were real. The county sheriff was also convinced.

One of the last of the Virginia gassings was reported in nearby Roanoke County. Afterward, the victim found discolored snow with a sweet-smelling, oily substance in it. When studied, it turned out to be a mixture of sulfur, arsenic, and mineral oil—insecticide ingredients. A trail of footprints led from the house to the barn, but none were found leaving the barn. They were, according to press accounts, a "woman's tracks."

Michael T. Shoemaker, who investigated the episode in the 1980s, noted its many similarities to the later scare at Mattoon. "In both Mattoon and Botetourt," he observed, "the principal effects were the same: a sickeningly sweet odor, nausea, paralysis, facial swelling and unconsciousness. These effects were confirmed by doctors and, moreover, in both cases doctors smelled the gas. Both gassers made repeat attacks

The gas caused nausea, headaches, facial swelling, and tightening of the mouth and throat muscles.

on one family, multiple attacks in one night and assaults on unoccupied houses. The pattern of explanation was also similar, progressing from pranksters to lunatics to hysteria. Tantalizing but useless clues were found," including *"a woman's print beneath a window."*

Gas attacks are still reported from time to time, typically in one building, such as a school, a factory, or a theater. For example, in March 1972 workers in a midwestern data-processing center complained of a mysterious odor that made them sick. Air, blood, and urine samples failed to detect anything out of the ordinary. When scientists who investigated the case eventually gave workers a false explanation—that an "atmospheric inversion" was responsible—the attacks of illness stopped!

Sources:

Coleman, Loren, *Mysterious America,* Boston: Faber and Faber, 1983.
Johnson, Donald M., "The 'Phantom Anesthetist' of Mattoon: A Field Study of Mass Hysteria," *Journal of Abnormal and Social Psychology* 40, 1945, pp. 175-186.
Shoemaker, Michael T., "The Mad Gasser of Botetourt," *Fate* 38,6, June, 1985, pp. 62-68.

FURTHER INVESTIGATIONS

BOOKS

Alien Contacts

Adamski, George, *Inside the Space Ships,* New York: Abelard-Schuman, 1955.

In 1952 Adamski, a UFO writer and photographer of space ships, reported that he met a flying saucer pilot from Venus. He then embarked on a colorful career as a contactee with connections on Mars, Venus, and Saturn. In 1954 a Venusian scoutcraft allegedly flew Adamski around the moon, and this book details his lunar odyssey.

Hopkins, Budd, *Intruders: The Incredible Visitations at Copley Woods,* New York: Random House, 1987.

Hopkins is best known for his UFO-abduction reports, which he, more than any other writer or investigator, has brought to wide public attention. *Intruders,* like his 1981 *Missing Time,* recounts the stories (many evoked under hypnosis) of witnesses who were abducted by large-headed, gray-skinned humanoids.

Strieber, Whitley, *Communion: A True Story*, New York: Beach Tree/William Morrow, 1987.

The best-selling UFO book of all time recounts the author's experiences with "visitors"—small, almond-eyed, gray-skinned humanoid occupants of UFOs. A fairly well-known writer of Gothic and futuristic fiction, Strieber contacted UFO-abduction investigator Budd Hopkins after a strange but barely remembered alien contact experience. He wrote this book after hypnosis, which revealed several visitor-related events in his life. William Morrow paid

Strieber $1 million for the book, which attracted enormous attention and a huge reading audience. The film version, starring Christopher Walken as Strieber, met with a fairly tepid response.

Ancient Astronauts

Temple, Robert K. G., *The Sirius Mystery,* New York: St. Martin's Press, 1977.

If the ancient astronaut fad of the 1970s produced one book of substance, many agree this is the one. Learned and extensively researched, it presents a complex, many-sided argument for an early extraterrestrial presence in West Africa.

Von Däniken, Erich Anton, *Chariots of the Gods?: Unsolved Mysteries of the Past,* New York: G. P. Putnam's Sons, 1970.

Along with several other writers in the 1960s, Swiss writer von Däniken theorized in this best-selling book that the gods of Judaism, Christianity, and other religions were extraterrestrials who, through direct interbreeding with our primitive ancestors or through direct manipulation of the genetic code, created *Homo sapiens.* These beings were also responsible for the archaeological and engineering wonders of the ancient world, as well as the mysterious Nazca lines. By no means the original text on the ancient astronaut theory, von Däniken's book took the world by storm, spawning a multitude of sequels, other books, and a popular film on the topic.

Anomalies, general

Fort, Charles, *The Books of Charles Fort,* New York: Henry Holt and Company, 1941.

Until Fort, the pioneer of unexplained physical phenomena, began his extensive research into anomalies, no one knew how "ordinary" and frequent strange happenings really were. This 1941 collection contains *Book of the Damned* (1919), a highly celebrated book that first exposed the reading public to giant hailstones, red and black rains, falls from the sky, unidentified flying objects, and other anomalies. Also in the volume are *New Lands* (1923), *Lo!* (1931), and *Wild Talents* (1932). Along with collecting and recording anomalies, these books present Fort's famously outlandish "theories," which he satirically regarded as no less preposterous than those scientists were offering to explain anomalies.

For further investigation of general anomalies see:

Bord, Janet, and Colin Bord, *Unexplained Mysteries of the 20th Century,* Chicago: Contemporary Books, 1989.
Cohen, Daniel, *The Encyclopedia of the Strange,* New York: Dorset Press, 1985.
Coleman, Loren, *Curious Encounters: Phantom Trains, Spooky Spots and Other Mysterious Wonders,* Boston: Faber and Faber, 1985.

Corliss, William R., ed., *Handbook of Unusual Natural Phenomena,* Glen Arm, MD: The Sourcebook Project, 1977.

Knight, Damon, *Charles Fort: Prophet of the Unexplained,* Garden City, NY: Doubleday and Company, 1970.

Michell, John, and Robert J. M. Rickard, *Living Wonders: Mysteries and Curiosities of the Animal World,* London: Thames and Hudson, 1982.

Bermuda Triangle

Berlitz, Charles, with J. Manson Valentine, *The Bermuda Triangle,* Garden City, NY: Doubleday and Company, 1974.

The Bermuda Triangle fever peaked with the publication of this best-selling book, which sold over five million copies worldwide. Like most of the Triangle books, there is little evidence of original research in its account of the disappearances of planes and boats off the Florida coast, and many of the "facts" that created the mystery were later discredited.

Kusche, Lawrence David, *The Bermuda Triangle Mystery—Solved,* New York: Harper & Row, 1975.

A thorough debunking of what Kusche calls the "manufactured mystery" of the Bermuda Triangle disappearances. For this book, Kusche did the research other Triangle writers had neglected. Weather records, newspaper accounts, official investigators' reports, and other documents indicated that previous Triangle writers had played fast and loose with the evidence.

Cattle Mutilations

Kagan, Daniel, and Ian Summers, *Mute Evidence,* New York: Bantam Books, 1984.

The authors traveled extensively through the western United States and Canada researching the bizarre stories of cattle mutilations. This definitive account exposes journalistic sensationalism and mass hysteria as the only solid basis for the modern-day myth.

Crop Circles

Delgado, Pat, and Colin Andrews, *Circular Evidence,* London: Bloomsbury, 1989.

Delgado, Pat, and Colin Andrews, *Crop Circles: The Latest Evidence,* London: Bloomsbury, 1990.

Two best-selling books on the English phenomenon. The authors, who are active crop circle investigators and founders of the Circles Phenomenon Research group, do not believe that the scientific explanations offered for the mystery so far—either weather- or hoax-related—can begin to explain the anomaly.

Cryptozoology

Bord, Janet, and Colin Bord, *Alien Animals,* Harrisburg, PA: Stackpole Books, 1981.

The best of the Bords' books on paranormal (supernatural) cryptozoology, *Alien Animals* expresses the authors' theory that all mysterious animal sightings, along with UFO sightings and other unexplainable apparitions, are manifestations of "a single phenomenon."

Heuvelmans, Bernard, *On the Track of Unknown Animals,* New York: Hill and Wang, 1958.

Heuvelmans, who is considered the father of cryptozoology, collected innumerable printed references to mysterious, unknown animals from scientific, travel, and popular literature and put them together in this large, informative book that sold over a million copies around the world.

Michell, John, and Robert J. M. Rickard, *Living Wonders: Mysteries and Curiosities of the Animal World,* New York: Thames and Hudson, 1982.

In this lively and literate book, Rickard, founder of the *Fortean Times,* and Michell capture Charles Fort's rich humor and sense of cosmic comedy, while providing encyclopedic, world-ranging coverage of current and historic anomalies.

For further investigation of cryptozoology see:

Caras, Roger A., *Dangerous to Man: The Definitive Story of Wildlife's Reputed Dangers,* New York: Holt, Rinehart and Winston, 1975.

Clark, Jerome, and Loren Coleman, *Creatures of the Outer Edge,* New York: Warner Books, 1978.

Mackal, Roy P., *Searching for Hidden Animals: An Inquiry into Zoological Mysteries,* Garden City, NY: Doubleday and Company, 1980.

Skuker, Karl P. N., *Mystery Cats of the World: From Blue Tigers to Exmoor Beasts,* London; Robert Hale, 1989.

South, Malcolm, ed., *Mythical and Fabulous Creatures: A Source Book and Research Guide,* New York: Greenwood Press, 1987.

Extinct Animal Sightings

Mackal, Roy P., *A Living Dinosaur?: In Search of Mokele Mbembe,* New York: E. J. Brill, 1987.

After two expeditions to the Congo to investigate reports of the mysterious mokele mbembe, University of Chicago biologist Roy Mackal persuasively argues in this book that the monster so frequently reported in this remote area of Africa is in fact some form of sauropod, a dinosaur thought to have been extinct for millions of years.

For further investigation of extinct animal sightings see:

Doyle, Sir Arthur Conan, *The Lost World* (fiction), London: Hodder and Stoughton, 1912.

Guiler, Eric R., *Thylacine: The Tragedy of the Tasmanian Tiger,* Oxford, England: Oxford University Press, 1985.

Folklore

Benwell, Gwen, and Arthur Waugh, *Sea Enchantress: The Tale of the Mermaid and Her Kin,* **New York: The Citadel Press, 1965.**

This highly regarded book on merfolk lore examines the many traditions of merfolk sightings throughout history and concludes that merfolk must be some kind of unknown, unrecorded species of sea animal.

Evans-Wentz, W. Y., *The Fairy-Faith in Celtic Countries,* **New York: University Books, 1966.**

The author, an anthropologist of religion, traveled throughout the British Isles recording oral traditions of fairy belief. The resulting book, a classic in folklore studies, also presents the author's extensive research on the existence of "such invisible intelligences as gods, genii, daemons, all kinds of true fairies, and disembodied men."

For further investigation of folklore see:

Otten, Charlotte F., ed. *A Lycanthropy Reader: Werewolves in Western Culture,* New York: Dorset Press, 1986.

Government Cover-ups

Moore, William L., with Charles Berlitz, *The Philadelphia Experiment: Project Invisibility—An Account of a Search for a Secret Navy Wartime Project That May Have Succeeded—Too Well,* **New York: Grosset and Dunlap, 1979.**

A popular book about the bizarre Philadelphia Experiment during World War II when, according to the letters of a very questionable character, a ship was made invisible and instantaneously transported between two docks, causing its crew members to become insane. The unflagging interest of three Office of Naval Research officers with the far-fetched story was the most unexplainable mystery here. Moore's book inspired the 1984 science fiction movie.

Randle, Kevin D., and Donald R. Schmitt, *UFO Crash at Roswell,* **New York: Avon Books, 1991.**

The government cover-up of the 1947 "Roswell incident" was effective enough to submerge the story of the crashed flying saucer for nearly 30 years, but the investigation is not closed. Randle and Schmitt have joined others in collecting the testimony of hundreds of witnesses, from local ranchers to air force generals. From these reports they have reconstructed a complex series of events, and the Roswell incident has become one of the best-documented cases in UFO history.

Hairy Bipeds

Napier, John, *Bigfoot: The Yeti and Sasquatch in Myth and Reality,* **New York: E. P. Dutton and Company, 1973.**

A scientist's view of the abundant evidence of Bigfoot's existence. Napier, a primatologist and the curator of the primate collections at the Smithsonian Institution, was one of the very few conventional scientists to pay serious attention to Bigfoot and other hairy bipeds.

Sanderson, Ivan T., *Abominable Snowmen: Legend Come to Life*, Philadelphia, PA: Chilton Book Company, 1961.

The first book to discuss Bigfoot/Sasquatch in any comprehensive manner. Sanderson linked the North American sightings with worldwide reports of "wild men," Almas, and yeti. An encyclopedic view of hairy biped traditions.

For further investigation of hairy bipeds see:

Bord, Janet, and Colin Bord, *The Bigfoot Casebook*, Harrisburg, PA: Stackpole Books, 1982.
Byrne, Peter, *The Search for Big Foot: Monster, Myth or Man?*, Washington, DC: Acropolis Books, 1976.

Lake Monsters

Zarzynski, Joseph W., *Champ: Beyond the Legend*, Port Henry, NY: Bannister Publications, 1984.

Vermont's own version of Nessie has a strong advocate in Zarzynski, who formed the Lake Champlain Phenomena Investigation in the 1970s for extensive research of historical sightings and surveillance of the lake. The author links Champ with the Loch Ness monster in this authoritative, if speculative, book.

Loch Ness Monsters

Dinsdale, Tim, *Loch Ness Monster*, 4th edition, Boston: Routledge and Kegan Paul, 1982.

Over a period of 27 years, British aeronautical engineer Tim Dinsdale made 56 separate expeditions to Ness and spent 580 days watching for the animals. In all, he had three sightings, one of which he filmed. The Dinsdale film is still considered compelling evidence of the existence of the monsters. His highly regarded book, *Loch Ness Monster,* went through four editions between 1961 and 1982.

Holiday, F. W., *The Dragon and the Disc: An Investigation into the Totally Fantastic*, New York: W. W. Norton and Company, 1973.

The author, the most radical of the Nessie theorists, originally suggested that the animals in Loch Ness were enormous prehistoric slugs. In this book he changed to an explicitly occult interpretation: Nessies are dragons in the most literal, traditional sense—they are supernatural and evil.

Mackal, Roy P., *The Monsters of Loch Ness*, Chicago: The Swallow Press, 1976.

Mackal was the scientific director of the Loch Ness Phenomena Investigation Bureau from 1965 to 1975 and this book, which grew out of his field work, is a cryptozoological classic.

For further investigation of Loch Ness monsters see:

Bauer, Henry H., *The Enigma of Loch Ness: Making Sense of a Mystery*, Urbana, IL: University of Illinois Press, 1986.
Binns, Ronald, *The Loch Ness Mystery Solved,* Buffalo, NY: Prometheus Books, 1984.

Sea Monsters

Heuvelmans, Bernard, *In the Wake of the Sea-Serpents,* **New York: Hill and Wang, 1968.**

In the most comprehensive volume ever written on the sea serpent, Heuvelmans analyzes 587 sea-serpent reports. He considers 358 of these to be authentic sightings, 49 hoaxes, 52 mistakes, and the rest lack sufficient detail to analyze. The author theorizes that the term "sea serpent" actually covers several unrecognized marine animals, which he classifies in the conclusive chapter.

Sanderson, Ivan T., *Invisible Residents: A Disquisition upon Certain Matters Maritime, and the Possibility of Intelligent Life under the Waters of This Earth,* **New York: World Publishing Company, 1970.**

Zoologist Sanderson demonstrates his wide-ranging curiosity and his creative imagination in this book in which he theorizes that an intelligent, technologically advanced civilization lives, undetected by the rest of us, in the oceans of the earth. These beings may be extraterrestrials, according to Sanderson, and some UFOs may be their versatile submarines.

For further investigation of sea monsters see:

Lester, Paul, *The Great Sea Serpent Controversy: A Cultural Study*, Birmingham, England: Protean Publications, 1984.

Unidentified Airships

Cohen, Daniel, *The Great Airship Mystery: A UFO of the 1890s,* **New York: Dodd, Mead, and Company, 1981.**

An entertaining, informative look at the famous turn-of-the-century UFO wave.

UFOs

Hynek, J. Allen, *The UFO Experience: A Scientific Inquiry,* **Chicago: Henry Regnery Company, 1972.**

At one time a consultant to the air force's UFO-debunking mission, astronomer Hynek changed from a skeptical attitude to a solid belief in the reality of UFOs. In his well-received book, *The UFO Experience,* he blasts the air force's UFO evidence-debunking projects and argues persuasively that science would be greatly furthered by an open-minded study of the subject.

Ruppelt, Edward J., *The Report on Unidentified Flying Objects,* **Garden City, NY: Doubleday and Company, 1956.**

When Lieutenant Ruppelt, an intelligence officer in the air force, took over the air force investigations of UFOs in the early 1950s, he insisted that investigations be carried out without prior judgments about the reality or unreality of UFOs. By the time he left the project two years later, Ruppelt was largely convinced that space visitors did exist. This memoir of his experiences is considered one of ufology's most important books.

Vallee, Jacques, *Passport to Magonia: From Folklore to Flying Saucers,* **Chicago: Henry Regnery Company, 1969.**

Vallee is one of the leading theorists on UFOs. *Passport to Magonia* marks his shift from scientific examination of UFO evidence to theories that UFO phenomena have their origins in another reality that is beyond the bounds of scientific analysis. Vallee proposes that inquirers need to immerse themselves in traditional supernatural beliefs—in fairies, gods, and other fabulous beings—in order to begin to understand that aliens and UFOs are only the modern manifestation of ancient beings.

For further investigation of UFOs see:

Blevins, David, *Almanac of UFO Organizations and Publications,* 2nd edition, San Bruno, California: Phaedra Enterprises, 1992.

Clark, Jerome, *The Emergence of a Phenomenon: UFOs from the Beginning through 1959—The UFO Encyclopedia,* Volume 2, Detroit, MI: Omnigraphics, 1992.

Clark, Jerome, *UFOs in the 1980s: The UFO Encyclopedia,* Volume 1, Detroit, MI: Apogee Books, 1990.

Weather Phenomena

Corliss, William R., ed., *Handbook of Unusual Natural Phenomena,* **Glen Arm, MD: The Sourcebook Project, 1977.**

Corliss, William R., ed., *Strange Phenomena,* **two volumes, Glen Arm, MD: The Sourcebook Project, 1974.**

Corliss, William R., ed., *Tornados, Dark Days, Anomalous Precipitation, and Related Weather Phenomena: A Catalog of Geophysical Anomalies,* **Glen Arm, MD: The Sourcebook Project, 1983.**

Corliss, a physicist who systematically catalogued more than 20 volumes' worth of anomalies, applies a conservative, scientific approach to bizarre and seemingly unaccountable events. He was particularly interested in unusual weather, and these volumes of his monumental Sourcebook Project are an invaluable resource in this area.

PERIODICALS

FATE

Llewellyn Worldwide, Ltd.
Box 64383
St. Paul, Minnesota 55164

Monthly

Fate was created in 1948 by Ray Palmer, science fiction editor of *Amazing Stories* and *Fantastic Adventures*, and Curtis Fuller, editor of *Flying*. Floods of flying saucer reports swept the nation following Kenneth Arnold's June 24, 1947, sighting of nine fast-moving discs. Palmer had found in his work that these kinds of "true mystery" stories were wildly popular, and, since no mass-circulation periodical devoted to such matters existed, the two editors decided to fill the void in the market.

Fate covered mysteries relating to ufology, cryptozoology, and archaeology, but its greatest emphasis was on psychic phenomena. In the 1950s Palmer sold his share of the magazine to Fuller, who then edited it with his wife, Mary Margaret Fuller. The magazine was the most successful popular psychic magazine of all time, and achieved a peak circulation of 175,000 during the late 1970s. In 1988 Mary Margaret Fuller was replaced as editor by Jerome Clark, and Phyllis Galde later took over. *Fate's* 500th issue was published in November 1991.

Strange Magazine

Box 2246
Rockville, Maryland 20852

Semiannual

In the introduction to the first issue of *Strange Magazine,* Mark Chorvinsky, the magician and filmmaker who created the publication, wrote: "We range from wild theoretical speculation to cautious skeptics—including every shade of worldview in between. Some of us are philosophers, others investigators and researchers—surrealistic scientists who catalog the anomalous, the excluded, the exceptional." The focus of *Strange* is on physical rather than psychic anomalies, and it includes topics such as cryptozoology, ufology, archaeological mysteries, falls from the sky, crop circles, and behavioral oddities. Chorvinsky, who has a particular interest in hoaxes, has exposed a number of dubious claims, most notably those associated with English magician and trickster Tony "Doc" Shiels, whose widely reproduced photographs of the Loch Ness monster and a Cornish sea serpent had many fooled. *Strange* reflects its editor's attitude toward anomalous phenomena: open-minded but not credulous.

Published semiannually, each issue of *Strange* is 64 pages long, full of lively graphics and well-written, well-researched articles. It is essential reading for all committed anomalists.

MAJOR ORGANIZATIONS
AND THEIR PUBLICATIONS

Ancient Astronaut Society

1821 St. Johns Avenue
Highland Park, IL 60035

Bimonthly bulletin: *Ancient Skies*

The Ancient Astronaut Society was formed in 1973 and is based on the belief that advanced space beings visited the earth early in human's history and possibly played a part in the development of human intelligence and technology. The organization is directed by attorney Gene M. Phillips in Chicago. European director Erich von Däniken, who wrote many books about the idea—including the wildly popular *Chariots of the Gods?*—operates out of Switzerland. The organization publishes a bimonthly bulletin, *Ancient Skies,* in both English and German. It also meets in a different world city every year and sponsors archaeological expeditions to sites where ancient marvels, viewed as evidence for the group's beliefs, can be seen by members firsthand.

Center for Scientfic Anomalies Research (CSAR)

Box 1052
Ann Arbor, MI 48106

Journal: *Zetetic Scholar*

Formed in 1981, the Center for Scientific Anomalies Research is a "private center which brings together scholars and researchers concerned with furthering responsible scientific inquiry into and evaluation of claims of anomalies and the paranormal." Director Marcello Truzzi and associate director Ron Westrum are sociologists of science at Eastern Michigan University in Ypsilanti. From 1978 to 1987 Truzzi, who had cofounded and then resigned from the Committee for the Scientific Investigation of Claims of the Paranormal (see CSICOP entry below), edited the journal *Zetetic Scholar,* a forum in which believers and nonbelievers could discuss and debate their views on anomalous subjects. CSAR was created as a parent organization for the publication; a number of important physical and biological scientists, psychologists, and philosophers are among its consultants.

Committee for the Scientific Investigation
of Claims of the Paranormal (CSICOP)

Box 703
Buffalo, NY 14226

Quarterly magazine: *Skeptical Inquirer*

The Committee for the Scientific Investigation of Claims of the Paranormal was formed in 1976 by Paul Kurtz, a professor of philosophy at the State University of New York at Buffalo, and science sociologist Marcello Truzzi (see CSAR entry above). Unfortunately, the two founders had different aims for the organization right from the start. Kurtz and his followers were rigid disbelievers in ufology, astrology, and other subjects that existed on the fringes of science; more than just skeptics, they viewed such unusual ideas as threats to reason and civilization. Truzzi felt that at least some claims, especially those made by serious parapsychologists, cryptozoologists, and ufologists, were worth investigating. Truzzi had hoped that the organization would practice fairminded *nonbelief* rather than ridiculing *disbelief.* He resigned a year later.

From the beginning CSICOP attracted many famous scientists. The organization was well funded and by the late 1980s its quarterly magazine, the *Skeptical Inquirer,* would claim a circulation of more than 30,000—the world's second most popular magazine on anomalies and the paranormal (after the psychic digest *Fate*). CSICOP sponsors regular conferences. Through its connected publishing house, Prometheus Books, it releases works expressing the debunker's view of UFOs, the Loch Ness monster, the Bermuda Triangle, and other "antiscientific" matters.

Fund for UFO Research, Inc. (FUFOR)

Box 277
Mount Rainier, MD 20712

The Fund for UFO Research was founded in 1979 and provides grants for scientific research and educational projects on UFO-related subjects. Studies have included investigations of UFO photographs and crash reports. The organization publishes findings from these projects from time to time. Funds are granted by a ten-member board of directors of scientists and scholars.

International Fortean Organization (INFO)

Box 367
Arlington, VA 22210

Quarterly journal: *INFO Journal*

The International Fortean Organization was founded in 1965 by Ronald J. Willis. It is dedicated to the memory and interests of pioneering anomaly collector Charles Fort (1874-1932). The *INFO Journal* is a forum for a wide variety of unexplained physical happenings, both past and present. The organization sponsors a yearly "FortFest" in the Washington, D.C., area, where well-known writers and researchers discuss anomalies.

INFO is a successor to the Fortean Society (1931-1960). Under the direction of writer and advertising man Tiffany Thayer, that organization and its *Fortean Society Magazine* (retitled *Doubt* in 1944) were known for their weird ideas and eccentric writers. The society oversaw the important publication of *The Books of Charles Fort,* a collection of the anomalist's writings, in 1941.

International Society of Cryptozoology (ISC)

Box 43070
Tucson, AZ 85733

Yearly journal: *Cryptozoology*

Quarterly newsletter: *ISC Newsletter*

The International Society of Cryptozoology was formed in early 1982. At its founding meeting at the Museum of Natural History of the Smithsonian Institution in Washington, D.C., the organization defined its purpose as the "scientific inquiry, education, and communication among people interested in animals of unexpected form or size, or unexpected occurrence in time and space." Roy P. Mackal, a University of Chicago biologist, and University of Arizona ecologist J. Richard Greenwell had worked behind the scenes for more than a year to put the organization together. (They, along with cryptozoology pioneer Bernard Heuvelmans, became the ISC's elected officers.)

With its serious, scientific approach to the subject of "unexpected" animals, the organization has been able to attract well-respected zoologists, anthropologists, and others as members. The ISC holds an annual meeting, always at a university or scientific institute. It publishes the quarterly *ISC Newsletter* and the yearly journal *Cryptozoology*. Although mainstream science has still not fully accepted cryptozoology, the ISC has enhanced the respectability of the field.

Intruders Foundation (IF)

Box 30233
New York, NY 10011

Yearly bulletin: *IF: The Bulletin of the Intruders Foundation*

The Intruders Foundation was created by Budd Hopkins, author of two popular books on UFO abductions. The purpose of the organization is to fund research and to offer therapeutic help to the many people who have contacted Hopkins, disturbed by their own claimed abduction experiences. The foundation has an informal national network of mental health professionals who volunteer to counsel these people. *IF: The Bulletin of the Intruders Foundation* appears yearly, and discusses abduction cases, investigations, and other related matters.

J. Allen Hynek Center for UFO Studies (CUFOS)

2457 West Peterson Avenue
Chicago, IL 60659

Journal: *Journal of UFO Studies* (*JUFOS*)

Newsletter: *International UFO Reporter* (*IUR*)

The Center for UFO Studies was formed in 1973 by J. Allen Hynek, the head of Northwestern University's astronomy department, and Sherman J. Larsen, a businessman and director of a small UFO group in Chicago. During the 1950s and 1960s Hynek had been the U.S. Air Force's chief scientific consultant on UFOs, until he publicly complained that the military was doing a very poor job of investigating reports. While the prevailing air force attitude toward UFOs was one of disbelief and dismissal, Hynek thought that UFOs were probably more than mistaken identifications and hoaxes. CUFOS was created so that scientists and other trained professionals could deal with UFO research in a serious but open-minded way.

CUFOS is one of two major UFO groups in the United States (the other is the Mutual UFO Network; see MUFON entry below). CUFOS publishes a newsletter, the *International UFO Reporter* (*IUR*), and the *Journal of UFO Studies* (*JUFOS*), and has sponsored UFO investigations. Located in Chicago, its huge collection of research materials is available to people studying UFOs. The organization's official name was expanded after Hynek's death in 1986.

Mutual UFO Network, Inc. (MUFON)

103 Oldtowne Road
Seguin, TX 78155

Monthly magazine: *MUFON UFO Journal*

Walter H. Andrus, Jr., a former officer of Tucson's Aerial Phenomena Research Organization, founded the Midwest UFO Network in 1969, based in Quincy, Illinois. In 1975 his group moved to Seguin, Texas, and became MUFON, the Mutual UFO Network. One of the most successful UFO organizations in the brief history of UFOs, it would claim 4,000 national and international members by 1992. Though open-minded about different explanations for UFOs, MUFON clearly leans towards the extraterrestrial hypothesis. The organization hosts a conference in a different U.S. city every year. MUFON's monthly magazine, *MUFON UFO Journal* (formerly *Skylook*), contains serious UFO studies and is essential reading for ufologists.

Society for the Investigation of the Unexplained (SITU)

Box 265
Little Silver, NJ 07739

Magazine: *Pursuit*

Founded in 1965 by science writer and lecturer Ivan T. Sanderson, the Society for the Investigation of the Unexplained publishes the magazine *Pursuit* about three or four times a year. *Pursuit* reports on anomalies and looks for the explanations or meaning behind them.

Society for Scientific Exploration (SSE)

Box 3818, University Station
Charlottesville, VA 22903

Semiannual journal: *Journal of Scientific Exploration*

Semiannual newsletter: *Explorer*

The Society for Scientific Exploration was formed in 1982. It encourages scientific study of UFOs, unexpected animals, supernatural claims, and other subjects that lie on the edges of science, because "progress towards an agreed understanding of such topics ... is likely to be achieved only if they are subject to the normal processes of open publication, debate, and criticism which constitute the lifeblood of science and scholarship." Full SSE members must be connected with a major university, government group, or corporate research institution; those who do not qualify are associate members. The society sponsors a conference each year at an American university. It publishes both the newsletter *Explorer* and the *Journal of Scientific Exploration* twice a year.

INDEX

Bold numerals indicate volume numbers.

A

I

Ice falls **1:** 137
Indian, American, anomaly reports
 2: 242-243; **3:** 440, 474, 476
Inorganic matter: falls from the sky
 1: 121
Inside the Space Ships **1:** 63
International Flying Saucer Bureau
 (IFSB) **1:** 48
International Fortean Organization
 1: 129, 177
International Society of Cryptozoology
 (ISC) **2:** 193, 197-199, 209, 271, 307,
 309, 343, 360; **3:** 387, 402, 439
International UFO Bureau **2:** 250,
In the Wake of the Sea-Serpents **2:** 197;
 3: 402
Intruders **1:** 15, 167
Intruders Foundation **1:** 15
Invisible Horizons **1:** 81, 94
Invisible Residents **1:** 96; **2:** 233; **3:** 379
In Witchbound Africa **2:** 312
Ishtar Gate **2:** 322, 324
Island of Lost Souls **2:** 219
It Came from Outer Space **1:** 14
Iumma **1:** 50
I Was a Teenage Werewolf **3:** 474

J

Jacko **2:** 254-257
Jaguar **2:** 341
J. Allen Hynek Center for UFO Studies
 (CUFOS) **1:** 14, 178-179
The Jessup Dimension **1:** 81
Jessup, Morris K. **1:** 26, 64, 80-83, 95,
 133, 141
Johnson, Robert **2:** 339
Jones, Mary **1:** 105-106
Journal of American Folklore **1:** 50
Journey to the Center of the Earth **1:** 57
A Journey to the Earth's Interior **1:** 56
Judaism **1:** 25
Jurassic Park **2:** 298

K

Kagan, Daniel **3:** 497-498, 517
Kangaroos, misplaced **2:** 203-205
Kaplan, Joseph **1:** 112-113
Kasantsev, A. **1:** 163
Keel, John A. **1:** 49-50, 59; **2:** 331, 333-335

Kelpies **3:** 410, 414
Keyhoe, Donald E. **1:** 9, 13-14, 64, 94, 96
King, Godfre Ray. *See* Ballard,
 Guy Warren
King Kong **2:** 239, 240, 301
Knight, Damon **1:** 133
Koldeway, Robert **2:** 324
Kongamato **2:** 312
Kraken **3:** 389
Krantz, Grover **2:** 232, 248
Kuban, Glen J. **2:** 325, 326
Kulik, Leonid **1:** 162
Kusche, Larry **1:** 93, 96-97, 98, 169

L

Lake Champlain **3:** 426-427
Lake Champlain Phenomena Investiga-
 tion (LCPI) **3:** 434
Lake monsters **3:** 409-444
Lake Okanagan **3:** 439-441
Lake Tele **2:** 308, 309
Lake Worth monster **2:** 252-254
The Land before Time **2:** 298
La Paz, Lincoln **1:** 111-113
Larsen, Sherman J. **1:** 8
Layne, N. Meade **1:** 58
Lazar, Robert Scott **1:** 85
Leaves: falls from the sky **1:** 131
Lee, Gloria **1:** 13
Legend of Boggy Creek **2:** 248
LeMay, Curtis **1:** 73
Lemuria **1:** 30, 57
Leonard, George H. **1:** 64
Leopard **2:** 206
Lescarbault **1:** 67-68
Leverrier, Urbain **1:** 66-69
Lewis and Clark expedition **1:** 160
Ley, Willy **2:** 196, 304, 322-324
Lights in folk tradition **1:** 103
Limbo of the Lost **1:** 96
Linnaeus, Carolus **2:** 194
Linnean Society **3:** 397-398
The Little Mermaid **3:** 465
"Little people" **1:** 26, 64, 82
A Living Dinosaur? **2:** 307
Living dinosaurs **2:** 197, 295-301
Living Mammals of the World **2:** 233
Lizards: entombed **2:** 225-226
Local lights **1:** 106
Loch Morar **3:** 434-438
Loch Ness **3:** 413-426
Loch Ness and Morar Project
 3: 423, 425

Q

Queensland tiger **2:** 213-214

R

Randle, Kevin **1:** 74
Red wolf **3:** 471
Regusters, Herman **2:** 296, 308-309
Religions **1:** 25
Religious lights **1:** 105-106
The Report on Unidentified Flying Objects (1956) **1:** 6
Reptile men **2:** 199, 329-331
Return of the Ape Man **2:** 239
Ri **2:** 198, 345-347
Ridpath, Ian **1:** 32-33
Road in the Sky **1:** 30
The Robertson Panel **1:** 6-7
Robins Air Force Base **1:** 5
Rocky Mountain Conference on UFO Investigation **1:** 47
Rogo, D. Scott **1:** 125
Rojcewicz, Peter M. **1:** 50
Roswell incident **1:** 14, 73-80, 171
The Roswell Incident **1:** 66
Rudkin, Ethel H. **2:** 338
Runaway clouds **1:** 150
Ruppelt, Edward J. **1:** 5, 6, 113

S

Sagan, Carl **1:** 32, 33
Salamanders: falls from the sky **1:** 131
Sananda **1:** 13
Sanderson, Ivan T. **1:** 81, 96, 98-99; **2:** 196-197, 233-234, 244, 255, 257-261, 267, 297-298, 304; **3:** 379
Sandoz, Mari **3:** 448
Sarbacher, Robert **1:** 78-79
Sasquatch **2:** 241-242
Satan worshippers **3:** 494-495
Saucer clouds **1:** 154
Saucer nests **3:** 503-504, 506
Saucers, flying **1:** 3-16
Sauropods **2:** 295, 302, 310, 323
Scandinavia **3:** 413
Schmidt, Franz Herrmann **2:** 298-300
Schmitt, Don **1:** 75
Science Frontiers **1:** 141
Scientific Study of Unidentified Flying Objects **1:** 9
Scully, Frank **1:** 75

Sea cows **3:** 426, 468
Sea monsters **3:** 375-380
Searching for Hidden Animals **2:** 307
Sea serpents **2:** 195; **3:** 394
Secret of the Ages **1:** 55, 57
Sedapa. *See* Orang-Pendek
Seeds: falls from the sky **1:** 131
Selkies **3:** 462
Serpents **2:** 322, 323; : 394-405; 447-452
Sesma, Fernando **1:** 50, 52
Seventh District Air Force Office of Special Investigations (AFOSI) **1:** 111
Shape-changing **2:** 337; **3:** 435, 473
Shaver, Richard Sharpe **1:** 57
Shiels, Anthony "Doc" **3:** 404, 421
Shooting stars **1:** 112
Siegmeister, Walter. *See* Bernard, Raymond
Sirius **1:** 31
Sirius B **1:** 31
Sirius mystery **1:** 31-33
The Sirius Mystery **1:** 31-33
Sirrush **2:** 322-324
"Sky gods" **1:** 30
Skyquakes **1:** 160-161
Sky serpents **3:** 447-452
Slick, Tom **2:** 268-270, 273
Smith, J. L. B. **3:** 449
Smith, John **3:** 466
Smithsonian Institution **2:** 193, 258-259
Smith, Wilbert B. **1:** 78
Society of Space Visitors **1:** 50
Solar system **1:** 68
Somebody Else Is on the Moon **1:** 64
Sonar **3:** 413, 425, 432
South Pole **1:** 57
Space animals **1:** 133; **2:** 340
Space brothers **1:** 12, 45-47
Space Review **1:** 48
Spencer, John Wallace **1:** 96
Sperm whales **3:** 391, 393
Spielberg, Steven **1:** 9, 79, 92
Splash **3:** 465
Spontaneous generation theory **1:** 121
Spontaneous human combustion **3:** 489-493
Sprinkle, R. Leo **1:** 47
Squids **3:** 388-393
Squid-whale battles **3:** 391-393
Star jelly **1:** 142-145
Starman **1:** 47
Steckling, Fred **1:** 63
St. Elmo's fire **1:** 156
Stewart, Jimmy **2:** 270